Health, Technology and Society

Series Editors
Rebecca Lynch
Life Sciences and Medicine
King's College London
London, UK

Martyn Pickersgill
Usher Institute
University of Edinburgh
Edinburgh, UK

Medicine, health care, and the wider social meaning and management of health are undergoing major changes. In part this reflects developments in science and technology, which enable new forms of diagnosis, treatment and delivery of health care. It also reflects changes in the locus of care and the social management of health. Locating technical developments in wider socio-economic and political processes, each book in the series discusses and critiques recent developments in health technologies in specific areas, drawing on a range of analyses provided by the social sciences. Some have a more theoretical focus, some a more applied focus but all draw on recent research by the authors. The series also looks toward the medium term in anticipating the likely configurations of health in advanced industrial society and does so comparatively, through exploring the globalization and internationalization of health.

More information about this series at
https://link.springer.com/bookseries/14875

Rhonda M. Shaw
Editor

Reproductive Citizenship

Technologies, Rights and Relationships

Editor
Rhonda M. Shaw
School of Social and Cultural Studies
Te Herenga Waka—Victoria University of Wellington
Wellington, New Zealand

Health, Technology and Society
ISBN 978-981-16-9450-9   ISBN 978-981-16-9451-6 (eBook)
https://doi.org/10.1007/978-981-16-9451-6

© The Editor(s) (if applicable) and The Author(s), under exclusive licence to Springer Nature Singapore Pte Ltd. 2022

This work is subject to copyright. All rights are solely and exclusively licensed by the Publisher, whether the whole or part of the material is concerned, specifically the rights of translation, reprinting, reuse of illustrations, recitation, broadcasting, reproduction on microfilms or in any other physical way, and transmission or information storage and retrieval, electronic adaptation, computer software, or by similar or dissimilar methodology now known or hereafter developed.

The use of general descriptive names, registered names, trademarks, service marks, etc. in this publication does not imply, even in the absence of a specific statement, that such names are exempt from the relevant protective laws and regulations and therefore free for general use.

The publisher, the authors, and the editors are safe to assume that the advice and information in this book are believed to be true and accurate at the date of publication. Neither the publisher nor the authors or the editors give a warranty, expressed or implied, with respect to the material contained herein or for any errors or omissions that may have been made. The publisher remains neutral with regard to jurisdictional claims in published maps and institutional affiliations.

Cover illustration: © shuoshu

This Palgrave Macmillan imprint is published by the registered company Springer Nature Singapore Pte Ltd.
The registered company address is: 152 Beach Road, #21-01/04 Gateway East, Singapore 189721, Singapore

Series Editors' Preface

Medicine, health care, and the wider social meanings and management of health are continually in the process of change. While the 'birth of the clinic' heralded the process through which health and illness became increasingly subject to the surveillance of medicine, for example, surveillance has become more complex, sophisticated, and targeted—as seen in the search for 'precision medicine' and now 'precision public health'. Both surveillance and health itself emerge as more provisional, uncertain, and risk-laden as a consequence, and we might also ask what now constitutes 'the clinic', how meaningful a concept of a clinic ultimately is, and where else might we now find (or not find) healthcare spaces and interventions.

Ongoing developments in science and technology are helping to enable and propel new forms of diagnosis, treatment, and the delivery of health care. In many contexts, these innovations both reflect and further contribute to changes in the locus of care and burden of responsibility for health. Genetics, informatics, imaging—to name but a few—are redefining collective and individual understandings of the body, health, and disease. At the same time, long established and even ostensibly mundane technologies and techniques can generate ripples in local discourse and practices as ideas about the nature and focus of healthcare shift in response to global debates about, for instance, One Health and Planetary Health.

The very technologies that (re)define health are also the means through which the individualisation of health care can occur—through, for instance, digital health, diagnostic tests, and the commodification of restorative tissue. This individualisation of health is both culturally derived and state-sponsored, as exemplified by the promotion of 'self-care'. These shifts are simultaneously welcomed and contested by professionals, patients, and wider publics. Hence they at once signal and instantiate wider societal ambivalences and divisions.

This Series explores these processes within and beyond the conventional domain of 'the clinic' and asks whether they amount to a qualitative shift in the social ordering and value of medicine and health. Locating technical use and developments in wider socio-economic and political processes, each book discusses and critiques the dynamics between health, technology, and society through a variety of specific cases, and drawing on a range of analyses provided by the social sciences.

The Series has already published more than 20 books that have explored many of these issues, drawing on novel, critical, and deeply informed research undertaken by their authors. In doing so, the books have shown how the boundaries between the three core dimensions that underpin the whole Series—health, technology, and society—are changing in fundamental ways.

In *Reproductive Citizenship*, Rhonda Shaw and colleagues explore medical and social infertility, the use of assisted reproductive technology (ARTs) and the biopolitical logics that surround these. By attending to structural infertility, this exciting contribution to the field assembles international research from a range of disciplines, forefronting largely under-represented voices in this area. Across themes of technologies, rights, and relationships, we find disruptions caused both to individual biographies and normative understandings of family and citizenship; as Shaw notes in her introduction, the 'good citizen' is 'not only economically productive, but familially (re)productive'. Through the book's lens of structural infertility, reproductive citizenship is further recast to include systems of privilege that enable and restrict the realisation of this as a goal. This timely work is not only an important contribution to the field of reproductive rights, obligations, hopes, and identities, but also the logics, ambivalences, and reverberations of health technologies and citizenship.

London, UK Rebecca Lynch
Edinburgh, UK Martyn Pickersgill

Acknowledgements

I am grateful to the authors whose research is featured in the following pages for their original contributions and commitment to this project. A special thanks to Emeline Robinson-Shaw and Ursula Robinson-Shaw for copy-editing and research assistance and to the Royal Society of New Zealand Marsden Fund and Te Herenga Waka—Victoria University of Wellington for supporting leave time for me to work on this collection. The collection has immensely benefitted from the generous and thoughtful suggestions of Katie Dow (University of Cambridge) and Rosemary Du Plessis (University of Canterbury).

Rhonda M. Shaw

Contents

1 Introduction: Reproductive Citizenship and Meanings of Infertility 1
   Rhonda M. Shaw

Part I Technologies 29

2 Affective Animacy and Temporalities in Danish Women's Accounts of Cryopreserved Embryos 31
   Michael Nebeling Petersen

3 The Affective Temporalities of Ovarian Tissue Freezing: Hopes, Fears, and the Folding of Embodied Time in Medical Fertility Preservation 51
   Anna Sofie Bach

4 Trans Narratives of Fertility Preservation: Constructing Experiential Expertise Through YouTube Vlogs 75
   Alex Ker and Rhonda M. Shaw

x Contents

5   Fertility and Fragility: Social Egg Freezing and the
    'Potentially Maternal' Subject   101
    Julie Stephens

Part II   Rights   125

6   Reproduction and Beyond: Imaginaries of Uterus
    Transplantation in the Light of Embodied Histories of
    Living Life Without a Uterus   127
    Lisa Guntram

7   Sized Out: Fatness, Fertility Care, and Reproductive
    Justice in Aotearoa New Zealand   153
    George Parker and Jade Le Grice

8   The Experience of Single Mothers by Choice Making
    Early Contact with Open-Identity or Private Sperm
    Donors and/or Donor Sibling Families in New Zealand   179
    Rochelle Trail and Sonja Goedeke

9   The Importance of a Genetic Link in Surrogacy
    Arrangements: Law, Public Opinion and Reconciling
    Conflict   203
    Debra Wilson

Part III   Relationships   227

10 Surrogacy and the Informal Rulebook for Making Kin
    Through Assisted Reproduction in Aotearoa New Zealand   229
    Hannah Gibson

11 Constructing Gay Fatherhood in Known Donor-Lesbian
    Reproduction: 'We get to live that life, we get to be parents' 253
    Nicola Surtees

12	**Doing Reflexivity in Research on Donor Conception: Examining Moments of Bonding and Becoming** *Giselle Newton*	279
13	**Reproductive Choices and Experiences in Planning for Parenthood and Managing Infertility** *Sonja Goedeke, Maria Mackintosh, and Lara Grace*	303

Index   329

Notes on Contributors

Anna Sofie Bach is a postdoctoral researcher in the Department of Sociology at the University of Copenhagen, Denmark. Her work cuts across the fields of family sociology, gender studies, and feminist science and technology studies. Anna Sofie's recent research comprises an ethnographic study of ovarian tissue freezing and transplantation, which has been published in *Science as Culture*, *New Genetics and Society* and *Reproductive Biomedicine Online*. The research is also part of the interdisciplinary volume *The Cryopolitics of Reproduction: A New Scandinavian Ice Age* (2020).

Hannah Gibson recently completed her PhD in Cultural Anthropology at Te Herenga Waka—Victoria University of Wellington, New Zealand. Hannah's research engages with the fields of science and technology studies, critical kinship studies, and medical and reproductive anthropology. Her PhD thesis, which explores how traditional and gestational surrogacy is practiced in Aotearoa New Zealand, involves ethnographic research with embryologists, lawyers, surrogates, and intended parents.

Sonja Goedeke is a senior lecturer at Auckland University of Technology, New Zealand. Sonja's research interests focus on the psychosocial and ethical implications of assisted reproductive technologies. She has published, presented, and supervised projects related to embryo donation, the concept of gifting in donation, counselling, mild stimulation protocols, policy/legislation, parenting/remaining childfree after IVF, timing of parenthood, embryo disposal decisions, and the recognition/compensation of donors.

Lara Grace lives in Auckland, New Zealand. She is an intern psychologist with a particular interest in women's health. Lara's research focuses on women's experiences subsequent to fertility treatment, including both of women who have had children following treatment and their adjustment to parenting, and of women who have not had children as a result of treatment.

Lisa Guntram holds a position as a senior lecturer in the Department of Thematic Studies and is affiliated with the Centre for Medical Humanities and Bioethics, Linköping University, Sweden. Lisa's research explores how embodiment, subjectivities, and normativities are shaped by, and shape, health care and medical practices. Her research has been published in *Social Science & Medicine*, *Feminist Theory*, *Bioethics*, and *Medical Humanities*.

Alex Ker is a social researcher and recent sociology graduate from Te Herenga Waka—Victoria University of Wellington, New Zealand. His research and community interests include the health and wellbeing of Rainbow young people and gender-affirming healthcare. He volunteers with InsideOUT Kōaro and is a member of the CCDHB's Sex and Gender Diverse Working Group.

Jade Le Grice is from the hapu Ngai Tupoto (Te Rarawa) and Ngati Korokoro, Ngati Wharara, Te Pouka, Te Mahurehure (Ngāpuhi). She is a senior lecturer in the School of Psychology at the University of Auckland, New Zealand. Jade's research focuses on reproductive decision-making and the interrelated domains of sexuality education, reproductive health, and abortion. Her work informs academic publications, psychology curriculum, and health policy.

Notes on Contributors   xv

Maria Mackintosh is a recent graduate of the Counselling Psychology programme at Auckland University of Technology, New Zealand. Maria has a particular interest in women's health and her Master of Art's research explored women's understandings regarding the 'right' timing for motherhood, focusing on factors that affect and constrain women's decision-making regarding parenthood.

Giselle Newton is a donor-conceived person and PhD candidate at the Centre for Social Research in Health at UNSW, Sydney, Australia. Giselle's doctoral research (2019–2022) explores the accounts of Australian donor-conceived people in online and offline environments, including their understandings of donor conception identities, communities, and support needs.

George Parker is a Strategic Advisor at Women's Health Action Trust and also lectures in the School of Health at Te Herenga Waka—Victoria University of Wellington, New Zealand. George recently completed a PhD in Sociology at the University of Auckland. A registered midwife, George has spent a number of years working in women's health policy and research. George's research includes critical perspectives on maternal obesity, rainbow inclusive maternity care, and reproductive justice.

Michael Nebeling Petersen is an associate professor in Gender Studies at the University of Copenhagen, Denmark. Michael's research centres on questions of culture, power, and identity, and the intersections between gender, sexuality, kinship, race, and nation. Michael has worked on transnational commercial surrogacy, focusing on how new reproductive technologies together with social media change cultural understandings of family, identity, and kinship, and has published widely in this area.

Rhonda M. Shaw is an associate professor in the School of Social and Cultural Studies at Te Herenga Waka—Victoria University of Wellington, New Zealand. Rhonda's research interests include the sociology of morality and ethics, empirical research on the donation and provision of human biological materials and services, and family life and intimate relationships.

Rhonda is a Principal Investigator on the Royal Society New Zealand Marsden Fund (2019–2023) project, *Accessing Assisted Reproduction: Social Infertility and Family Formation.*

Julie Stephens is an honorary professor at Victoria University in Melbourne, Australia. She is author of *Confronting Postmaternal Thinking: Feminism, Memory and Care* (2011). Her publications investigate areas of cultural activism and memory, changing meanings of the maternal, the social dimensions of mothering, feminist oral history, and the reshaping of emotions and care under neo-liberalism.

Nicola Surtees is a senior lecturer in the College of Education, Health and Human Development at the University of Canterbury, New Zealand. Nicola's teaching and research include a focus on family diversity; new forms of relatedness and processes of kin differentiation in the context of assisted reproduction; and curriculum, heteronormativity, social justice, and inclusion in early childhood education.

Rochelle Trail is a Single Mother by Choice of a donor-conceived son and as such has a special interest in the area of alternate family formation, particularly in the New Zealand context. Rochelle's research has focused on the experience of being a Choice Mother and the experience of recipients making early contact with sperm donors. She is also involved with *FertilityNZ* as the single women support coordinator.

Debra Wilson is Associate Professor of Law at the University of Canterbury, New Zealand, and a Principal Investigator of *Rethinking Surrogacy Laws*, a project funded by the New Zealand Law Foundation. Debra regularly speaks at international conferences on legal issues relating to surrogacy and has published many journal articles and book chapters on this topic.

1

Introduction: Reproductive Citizenship and Meanings of Infertility

Rhonda M. Shaw

Since the publication of T. H. Marshall's (1950) classic essay, *Citizenship and Social Class*, understandings of modern citizenship have emphasised civil, political, and social rights as well as corresponding responsibilities. While citizenship has been theorised in a variety of different ways, recent conceptions have challenged Marshall's three-stage classification to encompass broader understandings of citizen-subjects based on diverse claims to identity, inclusion, recognition, respect, and participation. Academics and researchers now refer to myriad ways of thinking about citizenship as an extension of the Marshallian perspective, evidenced by the ever-expanding vocabulary of citizenship couplings in the literature: bio-digital citizenship (Petersen et al., 2019), biological citizenship (Rose & Novas, 2005), cultural citizenship (Pakulski, 1997), everyday citizenship (Hopkins et al., 2015), women's citizenship (Lister, 1997), flexible citizenship (Ong, 1999), intimate citizenship (Plummer, 2003),

---

R. M. Shaw (✉)
School of Social and Cultural Studies, Te Herenga Waka—Victoria University of Wellington, Wellington, New Zealand
e-mail: rhonda.shaw@vuw.ac.nz

© The Author(s), under exclusive license to Springer Nature Singapore Pte Ltd. 2022
R. M. Shaw (ed.), *Reproductive Citizenship*, Health, Technology and Society, https://doi.org/10.1007/978-981-16-9451-6_1

multicultural citizenship (Kymlicka, 1995), sexual citizenship (Richardson, 2000), trans citizenship (Hines, 2009), and so forth.

In this volume we draw on aspects of this scholarship to focus on a range of intersecting issues around reproductive citizenship in relation to infertility and procreative and parental rights. In doing so, we acknowledge that the definition of reproductive citizenship and the protections it affords vary widely in different jurisdictions and geopolitical locations, and that scholars who draw on the concept of reproductive citizenship attribute to it a variety of meanings and valences, as outlined below.

Reproductive citizenship is a central focus of biopolitics (Foucault, 1990; Rabinow & Rose, 2006; Rose, 1999) and refers to the government and regulation of persons who are recognised and included as members of a community or nation-state by virtue of their capacity and entitlement to create or build a family. Coinciding with wider scholarship around citizenship studies, reproductive citizenship is connected to rights and responsibilities that include whether, how, when, and with whom to procreate, the right to intimate and personal life, and the right to create what Susan Golombok (2015) has called 'new families'. According to Golombok (2015, 3), these new family formations, which are a consequence of medical and social infertility, were non-existent or 'hidden from society' until the latter part of the twentieth century, and include families headed by lesbian, gay, bisexual, transgender, and queer (LGBTQ) parents, Single Mothers by Choice, and heterosexual families created by assisted reproductive technologies (ARTs).

The burgeoning of these families by the twenty-first century supports Bryan Turner's (2001, 2008) argument that reproductivity has become a central citizenship norm in contemporary western societies, alongside paid employment, and in some instances replacing military or public service, both of which have been historically connected with citizenship (Lister, 1997). As Turner (2001, 196; 2008, 46) explains, low fertility rates, particularly in liberal democracies in the Global North, have led states to promote 'the desirability of fertility and reproductivity as a foundation of social participation' and as a hallmark of full contribution to society. Since the non-reproductivity of medically and socially infertile people can now be notionally resolved by ARTs, Turner contends that the right to reproduce and parent is available to everyone, regardless of gender and sexuality.

As contemporary citizenship in many nation-states is said to be organised around nationality, kinship, and belonging to a family (Anderson, 1991), human reproduction and parenting is regarded as an important part of normative adult development and identity. Involuntary infertility thus presents a 'biographical disruption' (Bury, 1982) to people's lives, which Turner (2008, 52) says is damaging to their mental health and can lead, as researchers have extensively shown, to psychological suffering, identity crisis, and social exclusion (Greil et al., 2010; Shreffler et al., 2020; Tonkin, 2019; Ulrich & Weatherall, 2000; Verhaak et al., 2005; Whiteford & Gonzalez, 1995).

While assisted and conceptive technologies circumvent infertility to reconfigure possibilities for kin-making, thereby providing solutions for people experiencing involuntary childlessness, feminist and queer studies scholars remain ambivalent about their use. Not only do ARTs sanction normative ideas and practices, by transforming reproduction and parenting into a public duty; the deep desire to have genetically related children can overshadow alternative pathways to parenthood such as adoption, fostering, or step-parenting, reiterating understandings of kinship as based in biology. Even where assisted reproduction changes generational lines or deconstructs biological givens about who should or could become mothers and fathers—enabling older women, single women, lesbian women, gay men, transgender, and non-binary people to build families where otherwise they might not have had the opportunity—the extension of parenting rights remains a contentious topic based on assumptions about who is a good parent (Bell, 2010; Briggs, 2018; Golombok, 2015; Mohr & Herrmann, 2021; Murphy, 2012; Parker and Le Grice in this volume).

Ken Plummer (2003) alludes to this in his discussion on the rift within gay and lesbian politics over claims to intimate citizenship as it relates to the legal recognition of marriage equality and the right to create families of choice. As Plummer observes, increasing civil and social inclusion of previously marginalised gay and queer identities in many liberal democracies around the globe has resulted in a new form of homonormativity (Duggan, 2002), and this has come at a price. Assimilation within the heteronormative mainstream, on the basis of marriage and family values, has produced hegemonic forms of the 'good citizen' as not only economically productive, but familially (re)productive, marginalising those who do not fit the norm. For Plummer, aligning LGBTQ subjectivities with homonormativity constructs a particular modality of citizenship that re-privatises social life and represents a loss to the political agenda. Lee Edelman (2004) and Jack Halberstam (2005) have likewise questioned the desire to achieve conceptions of family that approximate heteronormative ideals, critiquing the idea of a reproductively responsible adult who follows an expected pathway to good citizenship. As well as reaffirming 'the time of reproduction' as hetero- and chrononormative (Freeman, 2010; Nebeling Petersen in this volume) by distinguishing parenting as the emblem of adulthood and maturity, a key theme in this strand of work is the extent to which the opportunity and right to utilise ARTs positions citizen-subjects as clients or consumers of reproductive services (Mamo, 2013).

A second body of work incorporating critique of responsible citizenship comes from feminist scholars debating the entanglement of fertility preservation technologies with neoliberal ideology (Bach in this volume; Baldwin, 2018, 2019; Carroll & Kroløkke, 2018; Jackson, 2017; Stephens in this volume). First developed for patients with cancer at risk of treatment-induced infertility, and then women concerned about age-related fertility decline (Albertini, 2019), anticipatory and preventive fertility preservation technologies are now increasingly used by transgender and non-binary people[1] before commencing hormone therapy and medical transition. Described by Casey McDonald et al. (2011) as a kind of reproductive revolution akin to the invention of the oral contraceptive, cryopreservation technologies proactively treat infertility before it happens (van de Wiel, 2020). They are represented in the media, the internet, and by the medical profession as liberating, particularly for women seeking to delay childbearing for reasons to do with career advancement or the absence of a committed partner. As such, feminist critics argue that the marketing and promotion of these technologies as a safeguard for potential parenthood relies on individuals' willingness and ability to actively shape their destiny through personal choice. Rather than leaving their reproductive futures to chance, responsible citizen-subjects are persuaded to use cryopreservation techniques to manage the risks associated with their diminishing fertility until they are better positioned to become 'good' mothers and parents (Baldwin, 2018). In doing so, these technologies connect the ideals of neoliberalism and responsible self-governance to new forms of reproductive citizenship.

In contrast to the view of infertility as a neoliberal consumer issue, Damien Riggs and Clemence Due's (2013) elaboration of reproductive citizenship, which examines media representations of Australian people undertaking offshore surrogacy arrangements, problematises infertility as a social disability. On the one hand, the state imposes social obligations to reproduce, yet disadvantages those unable to do so unless they can resolve the problem of their infertility themselves. Building on Turner's (2001) account, Riggs and Due (2013) argue that reproductive heterosex is identified as integral to citizenship and the norm against which other forms of family building are measured. As these researchers suggest, in societies where the heterosexual mode of reproduction is valorised as the 'natural' way to reproduce, individuals who lack this capacity are limited in their ability to contribute fully to society and are thus susceptible to 'reproductive vulnerability' during their fertility journey. The mode of reproductivity for those who experience medical or social infertility therefore impedes their access to citizenship and the social and civic benefits it affords. Not only is their perception of themselves as good reproductive citizens compromised by the inability to make babies via heterosex; their reproductive vulnerability is magnified by the emotional, psychological, and logistical risks associated with accessing reproductive assistance abroad. This includes, for example, navigating complex bureaucratic and legal processes involved in seeking citizenship for children born offshore through transnational surrogacy arrangements (see Deomampo, 2014).

However, as contributors to this volume show, reproductive vulnerability is not always cast in negative terms and may be framed as enabling by providers and recipients of reproductive materials and services.[2] In these narratives, the potential to contribute to intimate relationships and reconfigure kinship lines through donative acts (Gibson in this volume; Surtees in this volume) or to connect with others through online support networks (Ker and Shaw in this volume; Newton in this volume) can serve to collectivise individuals' experiences of reproductive provision and assistance, forming the basis for new communities and active political engagement. How people experience and talk about their positioning within biopolitics as reproductive citizens (or not) therefore depends on the conditions of their arrangements, where they are located, and their citizenship status.

While accessing ART provides options for people who experience medical and social infertility, the analytical utility of these classificatory categories is a key concern throughout this volume. Before introducing the individual chapters in the collection, it is useful to clarify the use of these terms.

Medical Infertility

Margarete Sandelowski and Sheryl de Lacey (2002, 43) have described infertility as an invention, having many faces (see also Bell, 2014; Greil et al., 2010, 2011; Johnson et al., 2014). Although there is no consensus about what infertility means, it is typically characterised as a complex medical phenomenon with wide-ranging causes that cannot be traced to a single aetiological origin. The World Health Organization (WHO, n.d.) defines medical infertility as a 'disease of the reproductive system defined by the failure to achieve a clinical pregnancy after 12 months or more of regular unprotected sexual intercourse'. A revised and expanded WHO definition, published in 2017, considered infertility as generating 'disability as an impairment of function' (Zegers-Hochschild et al., 2017). Most fertility clinic websites and specialists adopt WHO's classic clinical definition, diagnosing infertility as the inability to conceive after one year of unprotected sexual intercourse, or the inability to carry a live pregnancy to term (e.g. see Gillett, 2017).

The problem with defining infertility in this way is that it is implicitly framed as a planned event that affects cisgender women and heterosexual couples, potentially ignoring the experiences of cisgender men and LGBTQ people who form single or multi-parent families outside of the two-parent norm. Moreover, because classifying a disease is not always a clear-cut process, the interpretation of infertility rests on normative conceptions about what constitutes abnormal functioning, illness, and pathology. To illustrate, Luxi Nie and colleagues ask whether early onset menopause constitutes medical infertility or is the result of natural processes, and whether tubal infertility associated with sexually transmitted infections such as Chlamydia trachomatis is a medical condition or the outcome of previous sexual partner choices (Nie et al., 2012). To extend Nie and colleagues' point, fertility impairment caused by Chlamydia could be the result of a combination of factors such as conflicting sociocultural and religious values that prevent the disclosure of sexual and reproductive health concerns and poor access to primary healthcare. Ann Bell (2010, 634), who has done empirical research on fertility with women from different socioeconomic backgrounds in the United States, contends that infertility is treated differently by the medical establishment depending on whether women's past actions or choices align with dominant social norms and mores. Bell shows that women of low socioeconomic status are blamed for infertility associated with sexually transmitted infections, whereas middle-class women experiencing age-related infertility are 'empathised' with, thereby reinforcing a stratified system that values some groups' reproductive health and futures over others.

In some jurisdictions, such as Aotearoa New Zealand,[3] where commercialisation of ARTs is prohibited, eligibility for access to publicly funded fertility treatment requires the diagnosis of a medical condition or reason for infertility. To qualify for funding, patients must meet strict Clinical Priority Assessment Criteria (CPAC) (Gillett & Peek, 2001). The CPAC place limitations on who can access fertility treatment funding and thus pose specific challenges for prospective patients, clients, and consumers. Women must, for example, be under 39 years of age at referral, Body Mass Index (BMI) must be within the range of 18–32 kg/m2, patients must be non-smokers or ex-smokers for more than three months, and New Zealand citizens or eligible to reside in New Zealand for at least two years. The stated aim of the CPAC tool, which is to balance need against likely ability to benefit, is meant to ensure equity of access to limited health resources. Accordingly, the CPAC score is 'designed to offer treatment to deserving couples who have a low or zero chance of conceiving naturally and in whom treatment offers at least a reasonable chance of success' (Gillett, 2017, 184). Because access to public funding is based on medical criteria, it does not automatically include people who are socially infertile, unless they are also deemed medically infertile.

This model presents healthcare obstacles for people whose bodies, identities, and ways of living do not fit the CPAC eligibility profile. Transgender people, for instance, have the right to access information and services about fertility preservation as part of an informed consent model of healthcare in Aotearoa, as well as the right to use appropriate services should they choose to do so (Oliphant et al., 2018). Current rules, however, mean that only people who are medically infertile or undergoing procedures or treatments that will irreversibly harm their fertility meet the accepted criteria to access public funding. Fertility clinic specialists maintain that gender-affirming hormone therapy results in permanent infertility for transgender women, so they are eligible to receive public funding for sperm cryopreservation. Yet public funding excludes transgender men who seek to cryopreserve their ovarian eggs (unless they undergo an oophorectomy), as clinicians maintain they can get pregnant, without the need for treatment, by delaying or pausing hormone therapy. These diagnoses are based on definitions of sterility, for transgender women, as an irreversible physiological condition, and infertility, for transgender men, as a socially liminal state. By treating fertility and reproduction as a choice for transgender men, clinicians reinforce ideas about who is infertile and which forms of infertility are based on medical criteria. In this case, the strict definition of medical infertility not only means that transgender men have to pay out-of-pocket for egg cryopreservation; it also begs the question about the potentially negative effects for some individuals of interrupting hormone therapy to initiate fertility preservation or carry a pregnancy. 'Unlike cancer or heart disease', as Sandelowski and de Lacey (2002, 37) note, 'infertility still needs to be justified as worthy of public expenditure'.[4]

Social Infertility

Social infertility is commonly regarded as a voluntary fertility outcome. The term has become ubiquitous in the popular press and social media. In an early set of guidelines for infertility counselling, 'social' infertility was coined to identify a group of patients 'who seek fertility services because of their social circumstances rather than their medical status' (Boivin et al., 2001, 1304). Writing in the decade past, journalists and online bloggers have picked up on particular aspects of this usage. Amy Gray (2016) claims that social infertility is used in relation to delayed parenthood and older women seeking treatment; Armstrong (2016) and Melcantalk (2011) comment that it denotes the lack of a male partner during a woman's reproductive years. These definitions popularly refer to people who would like to have children but are single or experience age-related infertility and cannot reproduce without assistance. So, where the category of medical infertility is taken to refer to individuals who cannot physically reproduce via heterosex, this interpretation of social infertility refers to individuals who do not engage in heterosex 'at the right time'.

The Kennedy Institute for Ethics (2016) also emphasises the relational angle of social infertility, problematising medical definitions that automatically exclude lesbian and gay cisgender couples who are not having 'vaginal penile sex'. Accordingly, this expanded definition has been used to critique a wide range of issues around equity of access to ARTs for LGBTQ individuals and couples (see Lee et al., 2014; Lo & Campo-Engelstein, 2018; Liu, 2009; Peterson, 2005). Recently, Katherine Johnson et al. (2014, 29) have described the social circumstances relating to infertility in terms of situational fertility barriers. For people in these situations, infertility is the result of being 'busy with other life activities', as Bill Boddington and Robert Didham (2009) explain, and include a range of issues: economic pressures relating to limited employment opportunities, career, education pursuits, paying off a mortgage, re-partnering, changes to family make-up, being single, or being in a same-sex or gender diverse relationship. Lia Lombardi (2015), for instance, speaks of social infertility in Italy as a socio-political product, stemming from labour market precarity and unequal gender distribution of housework and childcare, leading to low fertility rates and delayed childbearing. In this rendering, social infertility, which is impacted by age-related factors, constitutes what I have elsewhere described as 'an outcome of life chances and circumstance' (Shaw, 2011, 1) and not a matter of personal preference or lifestyle choice as is often assumed.

The division between medical and social infertility is not self-evident, but rests on a false distinction that remains upheld in policy guidelines and practice in numerous jurisdictions in seemingly logical ways. Indeed, as Lucy van de Wiel argues, in relation to 'social' egg freezing for relationship or career reasons versus 'medical' egg freezing for chronic conditions such as cancer, the terminology 'polarizes a situation that is far more complex than this binary suggests' (2020, 38). In concert with van de Wiel and other feminist scholars who seek to disentangle the multiple meanings of infertility, the chapters in this volume question the medical-social dichotomy.

Structural Infertility

Although people prove to be remarkably inventive and resourceful in the face of fertility constraints (see Bell, 2010; Dempsey, 2016; Inhorn, 2005), access to ARTs remain compromised for many due to lack of health insurance, state funding, and legislative and policy barriers. Conventional beliefs about family formation relating to concerns about risk and harm to potential offspring and prospective mothers' or parents' capacity to provide satisfactory parental care compound these inequities. This reinforces 'stratified reproduction', the term coined by Shellee Colen (1995) to describe 'the power relations by which some categories of people are empowered to reproduce and nurture while others are devalued and even despised' (Van Balen & Inhorn, 2002, 16).

The oppositional framing of infertility as 'medical' versus 'social' further stratifies access to fertility treatment and assisted reproduction, by producing 'a particular image of *who* is infertile' and 'a particular understanding of *what* infertility is' (Bell, 2014, 2). As outlined above, social scientists have long critiqued the social construction of medical infertility as a discursive category, reflecting the bias of those making decisions about what biomedical criteria to include. Yet, while the criteria are ambiguous, Johnson et al. (2014) point out that some people, such as cisgender women, meet the definition of medical infertility if they intend to get pregnant but are unsuccessful. Where medical infertility is felt as a bodily impairment, this definition is a step towards recognising disadvantage for people who are trying to have children but cannot reproduce for biological or health-related reasons. Similarly, the concept of social infertility has proven useful to address the challenges people experience stemming from situational fertility barriers. While social infertility has been useful to describe people's relational identities, many people face fertility barriers not just because they are single, older, or LGBTQ, but because they experience compromised or denied access to ARTs as a result of discrimination and systemic inequalities. To shed light on the complexities of people's diverse experiences of reproductive vulnerability, a more robust concept than social infertility is needed.

Initially coined by Laura Briggs (2018, 16) in her discussion of reproductive labour and politics in the United States, the term structural infertility offers a way of moving beyond the limitations of the medical-social opposition. Rather than defining infertility as an individual medical problem or the result of a volitional act, Briggs characterises infertility as a condition with economic and social determinants. Much like Lombardi (2015), Briggs sees the development, marketing, and uptake of ARTs—of which egg freezing is a clear example—as part of a general process of neoliberal 'responsibilization' (Rose, 1999), resulting from long periods of financial insecurity, discriminatory employment, and gendered expectations around career, education, and later childbearing. Building on Briggs's argument about the stratification of reproductive labour in the global context of neoliberalism, the term structural infertility is used here to account for the ways in which access to ARTs and imagined future parenthood is socially stratified along gender, ethnicity, class, sexuality, dis/ability, and repronormative lines. As a lens through which to read the chapters in this volume, structural infertility denotes a system of privilege created by institutions, discourses, policies, and social practices that differentially enable and prevent individuals and groups of people from accessing and obtaining the resources needed to realise their goal of reproductive citizenship.

The Chapters in This Volume

Although ARTs are an increasingly global phenomenon,[5] they do not enter cultural voids or vacuums, but are shaped by local moralities, religious beliefs, and political values (Van Balen & Inhorn, 2002). Biopolitical discourses and organisational practices around ARTs in turn impact legislation and policy at a government and state level, having implications for citizens' rights and duties both onshore and globally. In this collection we include chapters by contributors that focus on assisted reproduction in the social, cultural, and geopolitical contexts of Australia, Aotearoa, Denmark, and Sweden, four liberal democracies that do not typically dominate centre-stage in the international reproductive studies literature, compared to research on ARTs from the United States, United Kingdom, and other parts of Europe.[6]

The Scandinavian countries of Denmark and Sweden are geographically worlds apart from the Antipodean nations of Aotearoa and Australia, and while each country has a distinctive configuration of citizenship rules and regulations, there are intercultural similarities in terms of affluence, fertility rates (1.6 to 1.7 births per woman), public healthcare, education, and welfare. In keeping with the national branding and image of all four countries as liberal and 'progressive' regarding inclusive legislation, same-sex marriage is legal, transgender individuals have access to fertility preservation options, and single and lesbian women have the right to access ARTs. This is not to say that access to fertility treatment was not previously restricted for some groups in these regions (e.g. see Millbank, 1997; Mohr & Koch, 2016).

With respect to ARTs, a quick snapshot of initial developments shows that Australia produced the first baby born from a donor egg in 1983 and one of the first births from a frozen embryo in 1984 (Marks et al., 2019); in the mid- to late 1980s, Aotearoa and Sweden were the first countries to introduce donor openness policies and recognise the child's right to biographical and genetic heritage information (Daniels & Lewis, 1996); currently, the state of Victoria in Australia has one of the most comprehensive donor-linking laws in the world (Dempsey et al., 2019); Denmark leads the world in the use of ARTs to build 'new families' (an estimated eight to ten per cent of all births) and is home to the largest sperm bank, Cryos International (Adrian, 2020; Proctor, 2018); and the Swedish uterine transplantation project was the first in the world to produce a successful birth from a live donor in 2014 (Guntram in this volume).

All four jurisdictions provide full or subsidised state funding for fertility treatment to those who qualify, and all permit altruistic gamete donation. All Australian states (except Northern Territory), Aotearoa, and Denmark prohibit commercial surrogacy contracts, but permit altruistic arrangements and allow reimbursement of 'reasonable expenses' or compensation. In Sweden, it is illegal for medical clinics to facilitate altruistic or commercial surrogacy, although intended parents from all four regions travel offshore for such arrangements. Like altruistic surrogacy in the domestic context, this invites discussion about claims to responsible citizenship that rely on others' gestational services to create non-normative families on the one hand yet reinforce inequalities on the other.

Our goal in this book is to bring together research contributions that examine the entangled biopolitical procreative and parenting logics that the people from these regions are engaged in, as well as offering timely discussions of new and emerging technologies. At the same time, we seek to centre the voices of individuals and groups who have been largely under-represented in the literature. To this end, the anthology is divided into 3 parts of 12 chapters. It includes contributions from 16 new and well-established researchers working in the fields of anthropology, bioethics, cultural studies, gender and sexuality studies, law, midwifery and health, psychology, science and technology studies, and sociology. While the chapters in the anthology clearly fall into categories based around specific technologies, such as gestational surrogacy and ovarian egg freezing, we have chosen to cluster them in relation to the subtitle of the book: technologies, rights, and relationships. Drawing on original, empirically oriented research, the essays examine a range of issues around various aspects of reproductive citizenship to do with conceptions of medical, social, and structural infertility.

Part I: Technologies

In the first section of the anthology, special attention is given to the ways in which various bio-technological options—notably gamete, embryo, and ovarian tissue freezing—are increasingly presented as ways of managing reproductive desire by deferring it to a future possibility. In Chap. 2, Michael Nebeling Petersen looks at the way cryopreservation in relation to ARTs is closely linked to temporality and kin. Petersen argues that the freezing of eggs and embryos is not only culturally interpreted as a social safeguard for reproductive futurity, but that it resolves anticipatory anxieties 'by maintaining the futurity of potential motherhood' (van de Wiel, 2020, 24). Based on 11 interviews with Danish women who have undergone fertility treatment to cryopreserve at least one embryo, the chapter poses that these women understand their frozen embryos not merely as reproductive matter presenting a potentiality, but as kin, constituting a paused (maybe collapsed) present. Drawing on theories around animacy hierarchies and queer temporalities, Nebeling Petersen discusses how the frozen reproductive matter of the embryo is affectively charged, given meaning, and animated as a result of its embeddedness within (heteronormative) kinship and (chrononormative) temporal orders.

Chapter 3, by Anna Sofie Bach, discusses ovarian tissue freezing as an emerging fertility preservation technique that is increasingly offered to patients whose fertility is compromised by either disease or medical treatment. Drawing on a combination of feminist science studies and queer phenomenology, Bach's chapter seeks to complicate the discussion about fertility preservation as a new type of hope-generating 'cryo-insurance'. Based on qualitative interviews with 42 Danish women, the chapter highlights the relation of hope and fear in preventive medicine and how embodied pasts, presents, and futures are constituted and (re)organised through cryopreservation and transplantation technology. While ovarian tissue freezing is clearly a future-orientated technology, Bach shows how it additionally extends the past into the present/future, in ways that oöcyte and embryo freezing does not. Bach suggests that frozen ovarian tissue transplantation also promises to restore the fertile, hormonal (feminine) body with the normalising rhythms of menstruation. In so doing,

1 Introduction: Reproductive Citizenship and Meanings... 15

the chapter demonstrates how connection to the pre-cancer body comes to work as a positive way for women to 'suture' pre- and post-cancer selves.

As cultural theorists and sociologists have shown, digital media technologies not only play an important role in online activism and the dissemination of information around health-related issues (Petersen et al., 2019); these platforms are also entangled 'with our daily lives, our bodily and intimate practices and our relationships' (Nebeling Petersen et al., 2018, 1). In the next chapter in the anthology, Alex Ker and Rhonda M. Shaw examine how intimacies intertwine with bodies and online technologies for transgender and non-binary people, who make video blogs (vlogs) about their fertility preservation journeys to share with others. The authors point out that transgender people currently find it difficult to access reliable information on fertility preservation options in mainstream settings, partly due to the misleading assumption that they are by default infertile or do not wish to have genetically related children. This lack of awareness is increasingly motivating transgender individuals to share their own views and experiences of fertility preservation through alternative media such as vlogging. Drawing on a virtual ethnography of YouTube vlogs, Ker and Shaw examine how the vloggers create and participate in epistemic online communities by constructing themselves as experiential experts through sharing their stories of, and views on, fertility preservation.

In the final chapter of this section, Julie Stephens maintains that while reproductive technologies were originally developed to deal with an urgent and immediate desire to have a child in the face of biological limits such as medical infertility and ageing, dominant discourses of fertility preservation now revolve around a particular neoliberal idea of choice. Stephens focuses specifically on the choice to harvest and preserve ovarian eggs for non-medical reasons in order for childbearing to occur in a future time when it is more professionally or personally convenient. She contends that 'the potential mother' has been produced to marketise cryopreservation technologies by reframing motherhood as an ever-deferred potentiality. This 'potential mother', says Stephens, is at the heart of one of the many contradictions of neoliberalism: the promotion of 'indefinite preservation' and the freezing of time, on the one hand, and directives towards incessant movement, career advancement, and change on the other. Stephens argues that the discourses surrounding these cryo-techniques represent a crisis in social reproduction, having implications for social and intimate relations, our understanding of care and human fragility, and ways of doing familial life.

Part II: Rights

Various social determinants limit people's access to ARTs, including financial barriers and the availability of accurate multilingual and culturally safe information. In this section, the authors focus on how the personal desire to become a parent, which Turner (2001, 2008) argues serves as a guarantee for full citizenship, motivates heterosexual and same-gender individuals and couples experiencing fertility issues to exercise their reproductive rights and choice.

The first contribution to this section, on uterus transplantation (UTx), bridges the chapters in Part I of the anthology. In this chapter, Lisa Guntram discusses UTx as a restorative reproductive technology that is believed to deliver what other alternatives to fertility treatment, such as surrogacy and adoption, cannot. Despite the risks associated with UTx, it may enable both genetic and gestational parenthood but is not considered successful until it has resulted in the birth of a healthy child. In the chapter, Guntram analyses in-depth interviews with women in Sweden who have pursued, or have considered pursuing, UTx to explore how their past experiences of living without a uterus, along with hopes of a future with a uterus, shape meanings and expectations accorded to UTx. Guntram focuses on aspects of the interviewees' accounts that are shaped by discourses of reproductive liberty, including the possibility to imagine new reproductive futures with UTx. At the same time as providing new choices and options for women, Guntram argues that we must also be attuned to the complexities of UTx-IVF procedures by questioning normative conceptions of gendered embodiment. Guntram concludes the chapter calling for an alternative basis for policy formulation around access and allocation that is more sensitive to the specificities of affected women's experiences and desires.

1 Introduction: Reproductive Citizenship and Meanings… 17

In the second chapter of the section, George Parker and Jade Le Grice examine equity of access to publicly funded fertility treatment in Aotearoa New Zealand. Their chapter focuses on five ethnically diverse women who were unable to obtain fertility care because they did not fit the required Clinical Priority Access Criteria (CPAC) stipulating that the BMI of the person seeking funding be less than 32 kg/m2. Clinical assessments of the use of CPAC to ration access to ARTs have affirmed the exclusion of women with high BMI, arguing that it incentivises 'compliance' with the weight loss recommendations given to people seeking fertility care and reduces the burden of so-called maternal obesity on maternity care. Parker and Le Grice present a critical discussion, operationalising the concept of structural infertility, to challenge the broader gendered, raced, and classed implications of these criteria in terms of who can access ARTs and form families. The authors conclude their chapter by drawing on principles of reproductive justice to insist on assisted reproductive technology policy (and funding) that decolonises ARTs and is fair, just, and inclusive.

Chapter 8, by Rochelle Trail and Sonja Goedeke, examines the experiences of Single Mothers by Choice (SMC) who make early contact with open-identity or private sperm donors and/or donor sibling families in Aotearoa New Zealand. Drawing on semi-structured interviews with Choice Mothers—a term adapted from Jane Mattes's 'SMC' organisation and Mikki Morrissette's 'Choice Moms' website—the chapter explores their motivations for seeking contact, their experiences of making contact, and how donor conception informs their family constructs. In Trail and Goedeke's analysis, the authors raise questions around challenges that arise from poorly understood expectations and boundaries; the complicated mix of family constructs made up of biological and emotional ties; and the matter of disclosure around donor conception. In line with the work of the New Zealand Law Commission, findings from the study support the need for further educational programmes facilitating recipients' disclosure to offspring, and the consideration of birth certificate annotation for donor-conceived people.

In Chap. 9, Debra Wilson examines legislation from the New Zealand Advisory Committee on Assisted Reproductive Technology (ACART), when considering an application for surrogacy involving assisted reproductive procedures, that at least one intending parent 'will be a genetic parent of any resulting child'. First discussing both the initial inclusion of the genetic link requirement under ACART Guidelines and the subsequent intention to remove it, Wilson then considers the results of a representative survey carried out in 2017–2018 to understand public perception of surrogacy and opinion on the importance of the genetic link. Wilson's analysis suggests that while many people will answer 'no' to the specific question of whether the requirement is necessary, answers to other questions suggest a difficulty in imagining why this link might not be present, thus reinforcing age-old assumptions that kinship should be fixed by biological relatedness as well as raising questions about who is legally the 'parent'. The upshot of the survey evidence Wilson examines both supports the ACART intended revision of its Guidelines, and indicates that the issue is not as simple as it might appear, highlighting the need for further discussion around parentage and surrogacy in Aotearoa New Zealand.

Part III: Relationships

Since the ground-breaking work of Mamo (2007), Stacey (1996), Weeks et al. (2001), and Weston (1991), a great deal of importance has been placed on contemporary familial life as increasingly chosen and fluid, generating long overdue questions 'about what being "family" and being related mean' (May & Nordqvist, 2019, 6). The third section of the anthology follows in this tradition. Drawing from various discourses of citizenship, the chapters investigate the affinities and links in people's lives as a consequence of using assisted reproduction, raising questions about the rights and interests of the parties involved to be recognised and included as legitimate reproductive citizens.

Non-commercial, altruistic surrogate pregnancy, which is permitted in numerous jurisdictions around the world, including Aotearoa New Zealand, is regarded as a legitimate family formation pathway for people experiencing medical and social infertility. In Aotearoa, traditional surrogacy can occur without fertility clinic intervention and ethical review, whereas gestational or clinic-assisted surrogacy is a regulated procedure under the legislation and must be approved via a formal ethics process. In Chap. 10, Hannah Gibson draws on three years of ethnographic fieldwork to explore how traditional and gestational surrogates and intended parents in Aotearoa manage the relationships involved in their reproductive journeys entailing surrogacy. With no social guide providing details on 'how to find a surrogate', Gibson draws on insights from international research to show that people ultimately model their search for surrogates and intended parents based on rituals and language associated with the dating world. From their search online for a partner to agreeing to go on a first 'date', relationships are carefully crafted with the understanding that there is a recipe for success handed down from more experienced members of the surrogacy community who provide a framework of conduct through explicit and implicit rules of engagement and advice. In the chapter, Gibson shows how emulating the practice of dating and adopting quasi-legal contracts in their interactions with one another provides the New Zealand surrogacy community with a road map to navigate feelings of reproductive vulnerability in an unknown situation.

Chapter 11, by Nicola Surtees, is based on a qualitative interview study undertaken in Aotearoa New Zealand investigating constructions of gay fatherhood in lesbian known donor reproduction. In the chapter, Surtees explores some of the ways in which gay couples exercise agency and choice, through their stories about plans for, or experiences of, sperm provision for lesbian couples and what they do to construct themselves as fathers and parents. Rather than approaching fatherhood as an entirely pre-given role that participants passively occupy, Surtees draws on the sociology of personal life (May & Nordqvist, 2019; Nordqvist, 2019) to examine what it means to create parenting identities as a set of activities and practices in relation to social norms around family building. In doing so, Surtees argues that prospective and established gay fathers/parents use available resources to account for their anticipated or actual fathering/parenting identities and practices, including biogenetic relatedness and statutory law to mitigate the vulnerability of their positions as third-party reproductive donors. Surtees concludes that the men narrate their self-as-father/self-as-parent identities through an existing mix of heteronormative tropes reflective of the dominant social order but that their self and identity construction work does challenge this order at times.

In the next chapter in the section, Giselle Newton draws on personal narrative to address the largely overlooked experiences of people who are donor-conceived. Newton focuses on the subject of donor conception by reflecting on her own positioning as a donor-conceived person, activist, and researcher. In the chapter, Newton draws attention to the fact that those for whom assisted reproduction was created in the first place—donor-conceived persons—have not been an integral part of academic and practical discussions on secrecy, anonymity, disclosure, and other aspects of assisted reproduction that affect their lives in fundamental ways. Because the lived experience of having been conceived through the use of donor gametes provides a unique perspective on the impact of ARTs, Newton says it is vital that research considers donor-conceived people as active agents and calls for greater involvement of donor-conceived people to influence services, policy, and legislation on this complex topic.

In spite of increased autonomy for women in reproductive decision-making and the trend to delay motherhood or choose to be childfree, the final chapter of the anthology shows how motherhood remains an expected stage of a woman's life in many cultures. At the same time, parenting expectations have changed across generations with the emergence in recent years of what has been termed a 'good mother' or 'new momism' discourse, encompassing a belief that parenting should be child-centred, that mothers are inherently 'better' parents than fathers, and that motherhood is characterised by fulfilment. In this chapter, Sonja Goedeke, Maria Mackintosh, and Lara Grace draw on research from three qualitative studies exploring women's experiences of reproductive vulnerability; one study examines young women's ideas about the appropriate timing for parenthood in Aotearoa New Zealand, and the others explore women's experiences after both successful and unsuccessful infertility treatment. The authors discuss how women's reproductive decision-making needs to be understood from a holistic perspective that acknowledges the biological parameters of fertility while also recognising social, cultural, and structural factors affecting women's experiences and influencing their reproductive autonomy. The chapter concludes by calling for a shift in thinking around how best to realise reproductive futurity, linking back to preceding chapters in the volume that address this question.

Ways of conceptualising citizenship in the twenty-first century are diverse and subject to ongoing critical debate. The following chapters in the volume contribute to this scholarship, drawing on empirically based research that examines claims to reproductive rights, obligations, hopes, and identities in a variety of contexts in which people experience medical, social, and structural infertility.

Acknowledgements I am grateful to Katie Dow for helpful comments and suggestions on an earlier draft of this chapter.

Notes

1. Transgender and non-binary people identify with a gender that is different to that which they were assigned at birth. While I use the umbrella term 'transgender' in this chapter to include transgender and non-binary genders, I acknowledge that not all people use nor relate to this term.
2. See Henk ten Have (2015) for a nuanced discussion of vulnerability.
3. Throughout this volume the terms Aotearoa, Aotearoa New Zealand, and New Zealand are used interchangeably. Aotearoa is a commonly used Māori name for New Zealand. Aotearoa New Zealand is used to acknowledge and honour Te Tiriti o Waitangi (the Treaty of Waitangi) as the founding document of the nation.
4. The decision not to fund egg freezing for transgender men in Aotearoa is endorsed by stakeholders as saving the public health dollar, since fertility clinic costs listed on local websites for oöcyte preservation can be as much as NZ$15,000 for egg collection, thawing, and embryo transfer. Sperm analysis and freezing by comparison is approximately NZ$400.
5. While access to ARTs in low-income countries remains limited (Inhorn & Patrizio, 2015), Hampshire and Simpson (2015) document research showing the increasing routinisation and acceptability of many ARTs in developing world settings and their availability to individuals and couples across diverse class and ethnic backgrounds.
6. There are anthologies including essays from contributors in these regions (e.g. see Kroløkke et al. (2015); Lie and Lykke (2016); Mackie et al. (2019); Nash (2014)), although empirical research from Aotearoa, in particular, has received relatively little international attention.

References

Adrian, S. W. (2020). Rethinking reproductive selection: Traveling transnationally for sperm. *BioSocieties, 15*, 532–554.
Albertini, D. F. (2019). Phase transitions in human ARTs: Fertility preservation comes of age. *Journal of Assisted Reproduction and Genetics, 36*, 1763–1765.
Anderson, B. (1991). *Imagined communities: Reflections on the origin and spread of nationalism*. Verso.
Armstrong, K. (2016). How the trend to delay parenthood leaves a gap in the lives of the older generation. Retrieved February 2018, from http://www.news.com.au/lifestyle/sunday-style/how-the-trend-to-delay-parenthood-leaves-a-gap-in-the-lives-of-the-older-generation/news-story/4bd601bb9b82d0d3211/
Baldwin, K. (2018). Conceptualising women's motivations for social egg freezing and experience of reproductive delay. *Sociology of Health & Illness, 40*(5), 859–873.
Baldwin, K. (2019). The biomedicalisation of reproductive ageing: Reproductive citizenship and the gendering of fertility risk. *Health, Risk & Society, 21*(5–6), 268–283.
Bell, A. V. (2010). Beyond (financial) accessibility: Inequalities within the medicalisation of infertility. *Sociology of Health & Illness, 32*(4), 631–646.
Bell, A. V. (2014). *Misconception: Social class and infertility in America*. Rutgers University Press.
Boddington, B., & Didham, R. (2009). Increases in childlessness in New Zealand. *Journal of Population Research, 26*, 131–151.
Boivin, J., Appleton, T. C., Baetens, P., Baron, J., Bitzer, J., Corrigan, E., et al. (2001). Guidelines for counselling in infertility: Outline version. *Human Reproduction, 16*(6), 1301–1304.
Briggs, L. (2018). *How all politics became reproductive politics: From welfare reform to foreclosure to Trump*. University of California Press.
Bury, M. (1982). Chronic illness as biographical disruption. *Sociology of Health & Illness, 4*(2), 167–182.
Carroll, K., & Kroløkke, C. (2018). Freezing for love: Enacting 'responsible' reproductive citizenship through egg freezing. *Culture, Health & Sexuality, 20*(9), 992–1005.
Colen, S. (1995). "Like a Mother to Them": Stratified reproduction and West Indian Childcare workers and employers in New York. In F. Ginsburg &

R. Rapp (Eds.), *Conceiving the new world order: The global politics of reproduction* (pp. 78–102). University of California Press.

Daniels, K., & Lewis, G. M. (1996). Openness of information in the use of donor gametes: Developments in New Zealand. *Journal of Reproductive and Infant Psychology, 14*(1), 57–68.

Dempsey, D. (2016). Relating across international borders: Gay men forming families through overseas surrogacy. In M. C. Inhorn, W. Chavkin, & J.-A. Navarro (Eds.), *Globalized fatherhood* (pp. 267–290). Berghahn Books.

Dempsey, D., Kelly, F., Horsfall, B., Hammarberg, K., Bourne, K., & Johnson, L. (2019). Applications to statutory donor registers in Victoria, Australia: Information sought and expectations of contact. *Reproductive Biomedicine & Society Online, 9*, 28–36.

Deomampo, D. (2014). Defining parents, making citizens: Nationality and citizenship in transnational surrogacy. *Medical Anthropology, 34*, 210–225.

Duggan, L. (2002). The new homonormativity: The sexual politics of neoliberalism. In R. Castronova & D. D. Nelson (Eds.), *Materializing democracy: Toward a revitalized cultural politics* (pp. 175–194). Duke University Press.

Edelman, L. (2004). *No future: Queer theory and the death drive*. Duke University Press.

Foucault, M. (1990). *The history of sexuality, volume 1: The will to knowledge* (Trans. Robert Hurley). Vintage Books.

Freeman, E. (2010). *Time binds: Queer temporalities, queer histories*. Duke University Press.

Gillett, W. R. (2017). Infertility. In C. Farquhar & H. Roberts (Eds.), *Introduction to obstetrics and gynaecology* (pp. 181–189). Nurture Foundation.

Gillett, W. R., & Peek, J. (2001). National specialist guidelines for investigation of infertility: Priority criteria for access to public funding of infertility treatment. Retrieved March 2018, from https://nsfl.health.govt.nz/system/files/documents/pages/infertility_cpac_tool.pdf/

Golombok, S. (2015). *Modern families: Parent and children in new family forms*. Cambridge University Press.

Gray, A. (2016). Social infertility: Can we blame women having IVF for delaying pregnancy? Retrieved February 2018, from http://www.abc.net.au/news/2016-05-31/social-infertility-blame-women-ivf-pregnancy/7461904/

Greil, A., McQuillan, J., & Slauson-Blevins, K. (2011). The social construction of infertility. *Sociology Compass, 5*(8), 736–746.

Greil, A. L., Slauson-Blevins, K., & McQuillan, J. (2010). The experience of infertility: A review of recent literature. *Sociology of Health & Illness, 32*(1), 140–162.

Halberstam, J. (2005). *In a queer time and place.* New York University Press.

Hampshire, K., & Simpson, B. (Eds.). (2015). *Assisted reproductive technologies in the third phase: Global encounters and emerging moral worlds.* Berghahn Books.

Hines, S. (2009). A pathway to diversity?: Human rights, citizenship and the politics of transgender. *Contemporary Politics, 15*(1), 87–102.

Hopkins, N., Reicher, S. D., & van Rijswijk, W. (2015). Everyday citizenship: Identity claims and their reception. *Journal of Social and Political Psychology, 3*(2), 84–106.

Inhorn, M. C. (2005). Religion and reproductive technologies: IVF and Gamete Donation in the Muslim world. *Anthropology Newsletter, 46,* 14–18.

Inhorn, M. C., & Patrizio, P. (2015). Infertility around the globe: New thinking on gender, reproductive technologies and global movements in the 21st century. *Human Reproductive Update, 21*(4), 411–426.

Jackson, E. (2017). The ambiguities of 'social' egg freezing and the challenges of informed consent. *BioSocieties, 13*(1), 21–40.

Johnson, K. M., McQuillan, J., Greil, A. L., & Shreffler, K. M. (2014). Towards a more inclusive framework for understanding fertility barriers. In M. Nash (Ed.), *Reframing reproduction: Conceiving gendered experiences* (pp. 23–38). Palgrave Macmillan.

Kennedy Institute for Ethics. (2016). Bioethics Blogs. Retrieved March 2021, from https://bioethics.georgetown.edu/2016/11/why-we-should-recognize-social-infertility-the-exclusion-of-lesbian-and-gay-couples-and-single-individuals-from-definitions-of-infertility/

Kroløkke, C., Myong, L., Adrian, S. W., & Tjørnhøj-Thomsen, T. (2015). *Critical kinship studies.* Rowman and Littlefield.

Kymlicka, W. (1995). *Multicultural citizenship: A liberal theory of minority rights.* Clarendon Press.

Lee, E., MacVarish, J., & Sheldon, S. (2014). Assessing child welfare under the Human Fertilisation and Embryology Act 2008: A case study in medicalisation. *Sociology of Health & Illness, 36*(4), 500–515.

Lie, M., & Lykke, N. (Eds.). (2016). *Assisted reproduction across borders: Feminist perspectives on normalizations, disruptions and transmissions.* Routledge.

Lister, R. (1997). *Citizenship: Feminist perspectives.* Palgrave Macmillan.

Liu, C. (2009). Restricting access to infertility services: What is a justified limitation on reproductive freedom? *Minnesota Journal of Law, Science and Technology, 10*(1), 291–324.

Lo, W., & Campo-Engelstein, L. (2018). Expanding the clinical definition of infertility to include socially infertile individuals and couples. In L. Campo-Engelstein & P. Burcher (Eds.), *Reproductive ethics II: New ideas and innovations* (pp. 71–83). Springer International Publishing AG.

Lombardi, L. (2015). Reproductive technologies and "social infertility" in Italy: Gender policy and inequality. In V. Kantsa, G. Zanini, & L. Papdopoulou (Eds.), *(In)fertile citizens: Anthropological and legal challenges of assisted reproduction technologies* (pp. 117–130). (In)FERCIT.

Mackie, V., Marks, N. J., & Ferber, S. (Eds.). (2019). *The reproductive industry: Intimate experiences and global processes*. Lexington Books.

Mamo, L. (2007). *Queering reproduction: Achieving pregnancy in the age of technoscience*. Duke University Press.

Mamo, L. (2013). Queering the fertility clinic. *The Journal of Medical Humanities, 34*, 227–239.

Marks, N. J., Mackie, V., & Ferber, S. (2019). Modes of mobility: Tracing the routes of reproductive travel in the Asia-Pacific region. In V. Mackie, N. J. Marks, & S. Ferber (Eds.), *The reproductive industry: Intimate experiences and global processes* (pp. 145–173). Lexington Books.

Marshall, T. H. (1950). *Citizenship and social class*. Cambridge University Press.

May, V., & Nordqvist, P. (2019). Introducing a sociology of personal life. In V. May & P. Nordqvist (Eds.), *Sociology of personal life* (2nd ed., pp. 1–15). Red Globe Press.

McDonald, C. A., Valluzo, L., Chuang, L., Poleshchuk, F., Copperman, A. B., & Barritt, J. (2011). Nitrogen vapor shipment of vitrified oocytes: Time for caution. *Fertility and Sterility, 95*(8), 2628–2630.

Melcantalk. (2011, February 3). 'Socially infertile'—A term that apparently refers to me and my fellow singletons. Retrieved February 2018, from https://melcantalk.wordpress.com/2011/02/03/socially-infertile-a-term-that-apparently-refers-to-me-and-my-fellow-singletons/

Millbank, J. (1997). Every sperm is sacred? *Alternative Law Journal, 22*(3), 126–129.

Mohr, S., & Herrmann, J. R. (2021). The politics of Danish IVF: Reproducing the nation by making parents through selective reproductive technologies. *BioSocieties*. https://doi.org/10.1057/s41292-020-00217-1

Mohr, S., & Koch, L. (2016). Transforming social contracts: The social and cultural history of IVF in Denmark. *Reproductive Biomedicine & Society Online, 2*, 88–96.

Murphy, T. F. (2012). The ethics of fertility preservation in transgender body modifications. *Bioethical Inquiry, 9*, 311–316.

Nash, M. (Ed.). (2014). *Reframing reproduction: Conceiving gendered experiences*. Palgrave Macmillan.

Nebeling Petersen, M., Harrison, K., Raun, T., & Andreassen, R. (2018). Introduction: Mediated intimacies. In R. Andreassen, M. Nebeling Petersen, K. Harrison, & T. Raun (Eds.), *Mediated intimacies: Connectivities, relationalities and proximities* (pp. 1–16). Routledge.

Nie, L., Anderson, L., & Henaghan, M. (2012). Surrogacy and s 2 (a) (ii): Interpreting the "medical condition" requirement in ACART's *Guidelines on surrogacy arrangements involving providers of fertility services* (2007). *New Zealand Family Law Journal*, March, 130–134.

Nordqvist, P. (2019). Un/familiar connections: On the relevance of a sociology of personal life for exploring egg and sperm donation. *Sociology of Health & Illness, 41*(3), 601–615.

Oliphant, J., Veale, J., Macdonald, J., Carroll, R., Johnson, R., Harte, M., Stephenson, C., & Bullock, J. (2018). *Guidelines for gender affirming healthcare for gender diverse and transgender children, young people and adults in Aotearoa New Zealand*. University of Waikato. https://researchcommons.waikato.ac.nz/handle/10289/12160

Ong, A. (1999). *Flexible citizenship: The cultural logic of transnationality*. Duke University Press.

Pakulski, J. (1997). Cultural citizenship. *Citizenship Studies, 1*(1), 73–86.

Petersen, A., Schermuly, A. C., & Anderson, A. (2019). The shifting politics of patient activism: From bio-sociality to bio-digital citizenship. *Health, 23*(4), 478–494.

Peterson, M. M. (2005). Assisted reproductive technologies and equity of access issues. *Journal of Medical Ethics, 31*, 280–285.

Plummer, K. (2003). *Intimate citizenship: Private decision and public dialogues*. University of Washington Press.

Proctor, L. (2018). Why is IVF so popular in Denmark? The changing face of procreation. *BBC World Service*. Retrieved September 27, 2021, from https://www.bbc.com/news/world-europe-45512312

Rabinow, P., & Rose, N. (2006). Biopower today. *BioSocieties, 1*, 195–217.

Richardson, D. (2000). Constructing sexual citizenship: Theorizing sexual rights. *Critical Social Policy, 20*(1), 105–135.

Riggs, D. W., & Due, C. (2013). Representations of reproductive citizenship and vulnerability in media reports of offshore surrogacy. *Citizenship Studies, 17*(8), 956–969.

Rose, N. (1999). *Powers of freedom: Reframing political thought.* Cambridge University Press.

Rose, N., & Novas, C. (2005). Biological citizenship. In A. Ong & S. J. Collier (Eds.), *Global assemblages: Technology, politics, and ethics as anthropological problems* (pp. 439–463). Blackwell Publishing.

Sandelowski, M., & De Lacey, S. (2002). The uses of a "disease": Infertility as rhetorical vehicle. In M. C. Inhorn & F. Van Balen (Eds.), *Infertility around the globe: New thinking on childlessness, gender, and reproductive technologies* (pp. 33–51). University of California Press.

Shaw, R. (2011). Infertility and childlessness. Te Ara—the Encyclopedia of New Zealand. Retrieved February 10, 2021, from https://teara.govt.nz/en/infertility-and-childlessness

Shreffler, K., Greil, A., Tiemeyer, S., & McQuillan, J. (2020). Is infertility resolution associated with a change in women's well-being? *Human Reproduction, 35*(3), 605–616.

Stacey, J. (1996). *In the name of the family: Rethinking family values in the postmodern age.* Beacon Press.

Ten Have, H. (2015). Respect for human vulnerability: The emergence of a new principle in bioethics. *Bioethical Inquiry, 12*, 395–408.

Tonkin, L. (2019). *Motherhood missed: Stories from women who are childless by circumstance.* Jessica Kingsley Publishers.

Turner, B. S. (2001). The erosion of citizenship. *The British Journal of Sociology, 52*(2), 189–209.

Turner, B. S. (2008). Citizenship, reproduction and the state: International marriage and human rights. *Citizenship Studies, 12*(1), 45–54.

Ulrich, M., & Weatherall, A. (2000). Motherhood and infertility: Viewing motherhood through the lens of infertility. *Feminism & Psychology, 10*(3), 323–336.

Van Balen, F., & Inhorn, M. C. (2002). Introduction: Interpreting infertility: A view for the social sciences. In M. C. Inhorn & F. Van Balen (Eds.), *Infertility around the globe: New thinking on childlessness, gender, and reproductive technologies* (pp. 3–32). University of California Press.

van de Wiel, L. (2020). *Freezing fertility: Oocyte cryopreservation and the gender politics of aging.* New York University Press.

Verhaak, C. M., Smeenk, J. M. J., Van Minnen, A., Kremer, J. A. M., & Kraaimaat, F. W. (2005). A longitudinal, prospective study on emotional adjustment before, during and after consecutive fertility treatment cycles. *Human Reproduction, 20*(8), 2253–2260.

Weeks, J., Heaphy, B., & Donovan, C. (2001). *Same sex intimacies: Families of choice and other life experiments.* Routledge.

Weston, K. (1991). *Families we choose: Lesbians, gays, kinship.* Columbia University Press.

Whiteford, L. M., & Gonzalez, L. (1995). Stigma: The hidden burden of infertility. *Social Science & Medicine, 40*(1), 27–36.

World Health Organization. (n.d.). Sexual and reproductive health: Infertility definitions and terminology. Retrieved February 2018, from http://www.who.int/reproductivehealth/topics/infertility/definitions/en/

Zegers-Hochschild, F., et al. (2017). The international glossary on infertility and fertility care, 2017. *Human Reproduction, 32*(9), 1786–1801.

Part I

Technologies

2

Affective Animacy and Temporalities in Danish Women's Accounts of Cryopreserved Embryos

Michael Nebeling Petersen

Introduction

An increasing number of people are turning to *in vitro fertilisation* (IVF) in order to become pregnant and have children. IVF treatment refers to the process of removing multiple oöcytes from a woman to be fertilised *in vitro* in the lab. The fertilised eggs are then grown for two to five days, after which one or two embryos are then inseminated within the uterus. If there is a surplus of embryos, these are cryopreserved and saved for later use.

In Denmark, IVF treatment and storage are included in universal health care, which means that the treatments are free of charge in public hospitals. In order to qualify for IVF treatments within the state healthcare system, a woman or couple will be assessed by a doctor to determine their suitability as parents. The assessment is done by the fertility doctor, and cases where the doctor cannot determine her/their suitability must

M. Nebeling Petersen (✉)
Gender Studies, University of Copenhagen, Copenhagen, Denmark
e-mail: nebeling@hum.ku.dk

be assessed by a national authority[1] (Mohr & Herrmann, 2021). Furthermore, the woman must be under 40 years and have a Body Mass Index (BMI) under 35; the woman must not already have a child, or the couple must not already be parents to a child together. Lesbian couples, single women, and heterosexual couples all must meet the same criteria. If the woman is older than 40 when starting treatment, or turns 41 during treatment, she will not be able to receive treatments in the public sector, but can use private clinics and pay for the treatments herself until she turns 45. After the age of 45, she cannot legally receive fertility treatment in Denmark. If a woman or a couple have surplus cryopreserved embryos from prior treatments, the woman or couple may use the embryos within the free and universal healthcare system until the age of 45, no matter how many children they have.

In Danish language, the words embryo and oöcyte are not normally used, but are referred to as fertilised and unfertilised eggs—often simply as eggs in political discourse. This conflation in language may be why the cryopreservation of both embryos and oöcytes is regulated in the same way in Denmark. These regulations make it illegal to store embryos (and oöcytes) for more than five years, and it is not possible to donate embryos to other women or couples. Instead, when the five years have passed, the woman or couple decide whether the cryopreserved embryo/s are destroyed or donated to science.[2]

Embryos have long been the subject of cultural anxieties, as they occupy a contested space of liminal life (Squier, 2004) or latent life (Radin, 2013); that is, should embryos be morally understood as lives, worthy of full protection? Are they just cells like any other human cells? Or are they something in between, a potential life? If so, is this potential life already genetically formed, and simply 'on hold'? Or is this genetic determinism a fantasy? These questions about the status of the embryo relate to questions about abortion and women's reproductive autonomy. It is therefore important to keep in mind that Denmark was one of the first Western nations to fully legalise abortion, and that every woman living in Denmark has the right to end a pregnancy of up to 12 weeks as part of free universal health care. Within political discourse and dominant public opinion, the right to free abortion is not questioned, and such questions would be commonly considered a political *faux pas*.

However, this does not mean that the moral status of the embryo is fixed; rather, the embryo is overdetermined, charged with many points of contestation, and as such, is open for the construction of new meanings (Laclau & Mouffe, 1985).

In this chapter, I will not attempt to find or develop a universal or moral status of the embryo, nor will I normatively discuss how an embryo could or should be understood. Rather, I want to understand how women who have cryopreserved embryos stored after fertility treatments relate to and understand these embryos. I will develop the analytical construct of 'affective animacy' as a way to understand how the embryos are animated; that is, how they are charged with feelings, meanings, and relations, and how these affects circulate.

Affective Animacy

I am interested in the ways in which the women obtain feelings for the embryos in relation to how affects circulate. In the stories of the women in the following study, they discuss the ways in which embryos take on different affective animations in different scenarios. This process, whereby an embryo is assigned new meaning and status as it disappears and reappears in different clinical settings, is what Charis Thompson has referred to as 'ontological choreography' (2005, p. 8). I suggest that these circulations or ontological choreographies can also be understood in terms of what Sara Ahmed calls 'affective economies' (2004). According to Ahmed, affect or feelings should not be conceptualised as something residing solely in an object or subject; rather, affect circulates among and within objects and subjects, constituting the borders and relations between them. These circulations, Ahmed suggests, can be conceptualised as economies, in a manner comparable to the conceptualisation of value in Marx's labour theory. The more an object circulates, the more affect it accrues; the more an affect circulates, the more value it obtains. As embryos circulate, appearing visually on screens and then becoming invisible to the eye, shifting from growing to static material when cryopreserved, they become objects for affective investments. These circulations charge the embryos with affect, and this affect is key to understanding how the embryos are animated in the experience of the women in my study. In her work on emotional choreographies, Stine Adrian has discussed the ways in which emotions shape and form reproductive ontologies in fertility clinics (Adrian, 2006). By introducing the concept of emotional choreographies (Adrian, 2006, p. 102), Adrian asks us to consider how the embryo is mediated via relations between human actors, non-human actors, norms, and discourses. I will examine how embryos are ontologically shaped by the complexities of these affective choreographies. While the universalised embryo is discursively dis-animated to 'merely' a collection of cells, I will consider how they are animated by the accrual of various affective values in the narratives of women who have undergone IVF.

Based on 11 interviews with Danish women who have undergone fertility treatments, and have at least one embryo cryopreserved, I analyse how the frozen reproductive matter of the embryo is affectively animated, given meaning, and awaken to life as a result of its embeddedness within (heteronormative) kinship and (chrononormative) temporal orders. On the one hand, the cryopreservation of embryos synchronises women to normative perceptions of family and temporality; on the other hand, cryopreservation disturbs normative conceptualisations of kinship, relatedness, and time. In this chapter, I argue that the women in this study domesticate the ('monstrous' potentialities of) cryo-technologies by *kinning* (Howell, 2006, p. 8; Kroløkke et al., 2016; Payne, 2016, p. 35) the frozen embryos and placing themselves and the embryos in chrononormative temporalities (Freeman, 2010), thereby culturally and affectively animating the cryopreserved embryos from frozen cells to living kin. As such, I suggest that frozen reproductive embryos are understood by these women not merely as reproductive matter, but as kin; not merely as potentiality, but as an intensified and already written futurity.

Methods of Data Collection

To collect data for this study, I conducted semi-structured interviews with 11 Danish women during 2018, all of whom at the time of the interviews had cryopreserved embryos stored. In all cases, the embryos were surplus embryos after in vitro fertilisation (IVF) treatments, and all but one woman had at least one child as a result of the treatments. The women all lived in Denmark; seven were from larger cities, three from small towns, and one was from a rural area. The women were aged 29–42; all had an academic or professional educational background; all had a steady job, though many were on maternity leave. The informants were all cisgendered, and identified as either heterosexual (nine) or lesbian (two). All interviewees were Danish citizens. Nine of the interviewees were white, one identified as Afro-Dane, and one was Danish-Asian. Seven of the women underwent the treatments while with a heterosexual partner, one of whom divorced during pregnancy. Two women had chosen to undergo treatment while unpartnered, and two women had IVF treatments and subsequently a child in a 'rainbow-family'; that is, with known fathers, who they did not have a romantic relationship to and who they did not live with.

All interviewees were recruited by a call for interviewees posted on Facebook. The call asked for women who had currently or in the past had their embryos cryopreserved and stored. The call was shared on my personal and open Facebook page, and reposted more than 20 times by friends and colleagues and their friends and colleagues (a sort of digital snowballing). Additionally, the call was posted from several fertility clinics' social media sites and on social media groups for people in fertility treatments. All interviews were conducted in a location chosen by the interviewee: most in their homes, and some in cafés close to their places of work. The interviews lasted 25–50 minutes and were based on the same interview guide, which focused on how the women felt about and gave a narrative account of their cryopreserved embryos. I started all interviews with an open question, asking the interviewee to explain how they ended up with embryos in the freezer, giving space for the women to speak freely with some clarifying follow-up questions. I then asked more directed questions about specific themes: for example, their experience of communication from clinics, their feelings about embryo destruction, and their reflections on the status of human embryos. Questions also included discussions about hypothetical embryo donation to other women, which is not possible in Denmark.

All interviewees have given informed consent to my use and publication of the data. All material is anonymised on two levels. Firstly, I have replaced all names, geographical places, and biographical markers, including workplace and education. Secondly, as the aim of this study is not to explore the life narratives of the women, but rather to examine the rhetorical strategies and affects which animate embryos and the meaning-making surrounding cryopreservation, I have chosen to present the interviews in a mixed format. I have divided the material according to the different dominant narratives that characterise them, so that the pseudonym of (for example) Karen is a composite of multiple interviewees. This is done to enable full anonymisation of the women, so that they cannot be readily identified by friends and partners.

Animacies and Strategy for Analysis

My analytical questions of how embryos are animated in the lives of the interviewed women are informed by the work by Mel Y. Chen's book *Animacies* (2012). In this book, Chen develops the linguistic concept of animacy—that is, how we use the language of agency and 'liveness' to differentiate animate/non-animate entities. Rather than thinking about life/death or human/non-human as fixed material categories, Chen develops the concept of *animacy hierarchies* to critically investigate how we linguistically and affectively animate entities in a life-hierarchy, where humans are at the top, stones at the bottom—and plants, animals, organic material, and indeed embryos, are somewhere in the middle. Chen combines this with a new materialist approach, arguing that processes of animacy deconstruct the binary between linguistic representation and matter by demonstrating how linguistic representation makes matter and how matter shapes linguistics. In this chapter, I use Chen's conceptualisation of animacy as a strategy of analysis to investigate how the interviewees linguistically, rhetorically, discursively, and affectively animate embryos.

In their study of how embryos are defined, envisaged, and imagined within debates about stem cell research, Williams et al. (2003) argued that the rhetorical representation of the embryo with the prefix human (e.g., 'human embryos'), as well as an rhetorical infantilisation of the embryo as an object to protect and keep safe, increased the humanisation of the embryos and worked as a rhetorical argument against the use of embryos in stem cell research, while simultaneously muting the voices of women. Cromer has shown how embryos are humanised rhetorically and racially (Cromer, 2019), and how embryo donation programmes fix the embryo in specific temporalities, which simplify the ontologies and potentialities of embryos (Cromer, 2018). In an older study, Franklin has discussed the cultural imaginaries that allow embryos to function as a 'salvation object' (1999, p. 64) for both stem cell researchers and 'pro-life' Christians. The framing and rhetorical positioning of the embryo enable it to emerge through multiple different ontologies (Roberts, 2007; Thompson, 2014). Thus, I do not understand embryos, or any other biological matter, as possessing given ontologies with universal status. Rather, I understand them as biological entities and emergent phenomena that are assembled, dis/animated, and accorded meaning via processes of power (Casper, 1998; Chen, 2012). Epistemologically, I am informed by the work on how reproduction is intimately embedded within histories of power and meaning-making; for example, colonial processes of racialisation and biological extraction (Vora, 2015), histories of regulating the female body by the medicalisation and disaggregation of pregnancy (Balsamo, 1995), and women's affective and discursive processes of embodiment, domestication (Berker et al., 2005), and meaning-making (Waldby, 2019).

Analyses

In the second section of this chapter, I will unfold my analyses. Firstly, I will discuss the narratives and meaning-making processes of the interviewees that serve to dis-animate and dehumanise embryos. I will then analyse how interviewees simultaneously animate embryos in processes of kinning and temporality.

Merely Cells?

All the interviewees positioned themselves as in favour of women's right to freely choose abortion, and when asked to what extent embryos should be considered living matter, they all answered that the embryos should be considered biological matter, but not living in a way that gives them any moral status. In Denmark, there is hardly any anti-abortion discourse, so this is not especially surprising. However, several of the women went on to describe 'their' cryopreserved as more than just cells. Mette told me:

> I don't think it is a child. Rather I think it is a precursor of a child. What I also keep thinking about is that the doctor, she said, 'it is a really strong and good blastocyst'. It wasn't one of those at the low scale, it was all the way at the top! It was the same with Valentin, he was also like that. I really feel it is a little gold egg laying up there [at the hospital]. When I cycle by the hospital, I usually wink to it.

Some women explained that these complicating feelings about the status of the embryos were due to difficulties related to assisted reproductive technologies (ART). For example, Karin told me that the feelings attached to 'her embryos' were due to many cycles of insemination, egg stimulation, and IVF that she and her partner had gone through before succeeding with this batch, from which she had one child, cryopreserving the surplus embryos. Similarly, other interviewees understood their feelings towards the embryos as a result of the ways in which clinicians and doctors talked about the embryos during the fertility treatments. Kirsten told me that the clinician, who had told how many oöcytes they took from her, had said that the oöcytes were 'pretty', and when the clinicians talked about how many embryos had evolved after IVF, they said that the embryos were 'strong and willful'. Kirsten explained: 'Who wants to destroy such beautiful little things? I know embryos cannot be willful, but still. I mean, they had made so far... They fought and won the odds. There must be something special about them...'. Kirsten's story shows how the cells are animated linguistically. The embryos are characterised by adjectives normally aligned with humans or living things, causing them to shift from object position to subject position. The embryos were described with agency, as when Kirsten says, 'They fought and won the odds.' In this way, Kirsten animates the embryo linguistically by framing some of the embryonic matter as 'heroic' matter. At the same time, this dis/animates the 'bad' matter, that is not 'strong and willful'. In this way, Kirsten animates some embryos while also dis/animating and consequently devaluing other embryonic matter.

As noted by Thompson, the categorisation of embryos as 'good' or 'bad' by medical staff in clinics and hospitals positions the embryos as either life or waste (2005, p. 265). In Karen's story, the categorisation of the embryo as 'strong' equates it characteristically with her already born child, who is also 'strong'. The metonymic connection between the laboratory categorisation of the embryo (now cryopreserved) and the characteristics of the child (already born) reminds us of how the embryos are further animated by the re-categorisation from simple cells to latent life in the clinic. As noted by Thompson (2005) and Chen (2012, p. 44), this animation and re-categorisation is biopolitically invoked by a more 'sensitized mapping of normalities'. The promise and latency of a normal future free from 'abnormalities' animate the embryonic matter with qualities of liveliness, whereas the presence of 'undesired conditions' de-animates the embryonic matter into waste.

Only one interviewee, Jane, did not have feelings for the embryo she had stored. Her feelings in relation to the fertility treatment were more to do with monetary economy. She was the only interviewee who did not end up having a child: her first fertilisation did not result in a pregnancy, the next two embryos were not successfully thawed, and before the remaining embryo was thawed and inseminated, she and her partner decided to stop the treatment. All the other women who expressed significant feelings towards the embryonic matter had at least one child resulting from the same batch of frozen embryos. When asked, the majority of interviewees expressed in detail their many thoughts, reflections, and emotions about the cryopreserved embryos. But Jane focused on the ways in which she felt she had been treated unfairly by the welfare system in relation to money: when she started treatment, she was 39, and had to rush in order to make sure that she had her treatments before turning 41. At the time of her interview, she was over 41, and if she wanted to have the last embryo inseminated, she would need to pay herself for services in a private clinic. In Jane's case, the donor of the sperm cell was a gay man, who had decided that he did not want to be a father anymore. Thus, Jane was left in a situation where her options were to have the embryo inseminated at her own cost or destroyed. I asked about how she felt about the embryo, which was stored in another city in Denmark. She replied that she had paid taxes her whole life, and she stressed that due to a high income, this amounted to quite a lot.

Jane did not in any way animate the embryo or discuss the embryonic matter, even though I prompted her to do so multiple times. Instead, she related my questions back to patient rights within the welfare system. In this discussion, the embryo was simply conceptualised as matter, not in terms of its animated potentiality. The placement of the embryo lower in the animacy hierarchy was related to its embeddedness in political and economic realms, rather than in the context of family and kinship.

Kinning of Embryos

Within media theory, Berker et al. (2005) argue that new technologies need to be 'domesticated': 'These "strange" and "wild" [media] technologies have to be "house-trained"; they have to be integrated into the structures, daily routines and values of users and their environments' (Berker et al., 2005; see also Møller & Nebeling Petersen, 2017; Silverstone, 1994). I use this line of thought to examine how the women domesticate the reproductive technologies and embryonic matters by integrating the '"strange" and "wild" technologies' into 'the structures, daily routines and values of users and their environments'.

When I interviewed Solveig, she told me about the status of her cryopreserved embryo: 'It is Laura's sibling, twin sister actually.' Laura is Solveig's daughter who is the result of an embryo from the same batch as the one cryopreserved. The sentiment, that the embryos were siblings to born children, was repeated in many of the interviews. Frequently, the cryopreserved embryo is described as part of a kinship relation via its imagined role as sibling to an actually existing son or daughter. By attaching the embryonic matter to a living child, the affective intensity of that parent-child bond attaches also to the embryo. We can understand this as a process of domestication, in which the ontological insecurity of the embryo is reduced. I call this process of domestication through affect *kinning of embryos*.

Interestingly, when the embryonic matter becomes part of the family, interviewees referred to it as a 'sister', 'brother', or 'sibling'. Positioning the embryonic matter within a familiar gendered and heteronormative script affectively animates the matter and simultaneously naturalises the (heterosexual) nuclear family. The linguist John Cherry (1992) has noticed how, across different languages, humans linguistically animate in hierarchies, animating adults over non-adults, males over females, higher/larger animals over smaller animals and insects, and familiar (kin/named) over unfamiliar (non-kin/unnamed). The naming of the embryo as part of a kin-relation places it highly in the animacy hierarchy.

In the following quote by Karen, she explains that it was only after the birth of her child that the embryonic matter became animated for her:

> I got pregnant and got Valentin and after that, the embryo has just haunted me! [...] I feel like this, when I look at Valentin, I often think, oh my, you have a twin brother or sister waiting for you. [...]

Notice, how the embryonic matter not only becomes animated through kinning, but also by becoming the subject to a transitive verb ('haunt') which has agency to do something to the grammatical object ('me'). In many ways, imagery surrounding genes and DNA (Franklin, 1999) contributes to the visualisation of the frozen embryo. Later in the interview, Karen continues:

> It is like instant noodles [...] I have like this, that it is a mini child laying there. It is like a little seed in a pot of earth, it just needs a little water, then it is a new life.

It is emphasised by many interviewees that it is the genetic fixity of the embryo (in contrast to an oöcyte) that makes it affectively important to them. As Karen explains, the embryo is in many ways fixed material matter, just like 'instant noodles', which only requires a simple process to become a fully fledged moral person ('just needs a little water'). In this way, genes are understood as the key and defining codes of human becoming. Expressed as such, the animation of the embryo is not solely discursive. Kinning of the embryo should be understood as an onto-epistemological process in which the matter of the cryopreserved embryo(s), the physicality of the born child, and the heteronormative imageries about family connectedness work together to affectively animate the embryo up the animacy hierarchy.

Temporal Synchronisation: Somatic and Generational Temporalities

Temporalities are embedded within timekeeping techniques and devices such as calendars and watches. These technologies inculcate 'hidden rhythms' (Zerubavel, 1981) to form temporal experiences that appear natural (e.g., 'my inner clock' (Freeman, 2010)) and work to orient people in specific ways (Adams et al., 2009). This means that the way time feels and the way it structures life becomes naturalised, organising bodies and identities according to normative scripts. In this line of thought, reproductive temporalities are not natural, given structures of life; rather, they are understood as effects of cultural hegemonies and capitalism (Halberstam, 2005; Zerubavel, 1981). The matter of how, and if, to have children, is thus embedded within cultural temporalities that are not biological, even if they appear or feel so (Kroløkke et al., 2020).

As argued by Kroløkke (2018), imageries about cryo-technologies work to synchronise reproductive potential (e.g., feelings of an 'inner clock') with ideals of romantic time (e.g., when to meet Mr. Right and when (not) to marry). This synchronisation is also prevalent in the narratives of the interviewees in this study.

> I was 37 years, when I started treatments. [...] I have always dreamed of becoming a mother, you know, always. In my thirties, I really felt my inner biological clock, and as I never met 'Mr. Right', God knows I have tried! So I decided to try to become a solo mother. [...] I never thought I would be a solo mother, but as time was running out, I guess it was natural for me to become it. (Interview with Lotte)

Lotte was 42 when I interviewed her. She explained that in her late thirties she had been *in sync* with gendered and reproductive temporal norms, and had structured her life to accommodate her wish to become a mother. She had chosen a career within education, which she believed was easy to adjust to a potential family life, and had left romantic relationships where she did not view the men as 'father potentials'. In her twenties, her life within a single culture in a larger city in Denmark appeared non-problematic, as heteronormative cultural imaginaries of a life span in Scandinavian late modernity include a rather long period of youth, stretching from late teens to early thirties, framed by liberal notions of coming of age with multiple partners. The dominant view is that in the early thirties, one should find a spouse, a steady job, and start a family.

When reaching her late thirties, Lotte's feeling of being in sync with normative expectations dissipated; her daily rhythms became more and more dissimilar to that of her friends, as they 'all had children during the same year or two'. This feeling of being out of sync also relates to the timescale of a 'biological clock', as Lotte began to feel an intense awareness of her decreasing fertility. The hidden rhythms of repro-time materialised for Lotte so much so that she *felt* her biology ticking inside of her. As she reached her late thirties, she felt that 'time was running out'. Entering ART and fertility treatments became a way of synchronising her somatic time and reproductive time to (hetero-)normative temporalities.

As shown by Waldby, for child-free women the cryopreserved oöcytes offer the possibility of 'a family ancestry and a potential line of descent' (2019, p. 8). In his book *In a Queer Time and Place* (2005) Jack Halberstam elaborates on temporalities in relation to reproduction. He explains how heteronormative temporalities are naturalised to govern and orient bodies and lives. He deconstructs reproductive chrononormativity (to borrow the concept of Freeman, 2010) in three temporalities: the time of reproduction (the 'biological clock', cultural norms about the appropriate age to marry and have children), family time (the everyday organisation of time in relation to the nuclear family, 'governed by an imagined set of children's needs' (Halberstam, 2005, p. 5)), and the time of inheritance (generational temporality and the time of family ancestry, which connects the family to its ancestors and nation).

In spite of difficulties, Lotte gave birth to a child before turning 40 years old: 'It was the happiest day of my life. I only regret that I didn't do it [become pregnant] earlier'. Through Lotte's temporal narratives, we can understand ART as a set of techniques that not only enable reproduction, but the synchronisation of (childless) bodies to both reproductive time (when (not) to have children) and family time (the daily rhythms of life).

Ten out of the eleven interviewees had already had a child as a result of IVF, and thus had already been lifted into generational time. In the experiences of these women, becoming a mother enabled them to move from a form of infantile temporality into generational time, which represents a connection to the family, but also immortalises the woman herself by rendering her a part of the (genetic) reproduction of family and nation. Thusly, the embryo itself represents the possibility of shifting parents from the state of infertility and its symbolic infantilism into a national temporality and coherence with the norms of nuclear family time.

Following David Eng (2010) and Rose (2007), Kroløkke and I argue in our analysis of the Swedish debates on commercial surrogacy and uterine transplants (2017, p. 205) that in Western late-capitalist societies, having a child serves as a guarantee for full and robust citizenship and for being recognised as a subject who has realised him- or herself economically, politically, and socially. In this way of conceptualising citizenship, it is less a question of nationality and more about becoming a responsible and active (reproductive) citizen. As such, we might understand the alignment to reproductive norms as a way to gain a cultural and political belonging within a chrononormative culture, in which the women move from a (temporal and affective) state of not-belonging to become citizens. According to Halberstam, having children is understood in dominant culture as a move from infantile temporality into generational time. Such a move enables women to synchronise with the chronology of inheritance and thereby connect to family ancestry and the nation. Thus, we can understand the women's temporal synchronisation as a turning to family time and part of what enables them to become (reproductive) citizens.

Further, we can understand the embryo as materialisation of temporality: the embryo holds the power to transform temporalities, and it is itself a sort of temporal potentiality. The embryo is not only animated by linguistic framing; the embryo animates back, holding a promise of a temporal synchronisation that elevates these women into normative reproductive temporalities of the body, the family, and the nation. It should come as no surprise, then, that the embryos are felt and understood as living kin, reaching out from the cryo-tanks into the temporal and affective lives of women.

Temporal Intensification

In Denmark, oöcytes and embryos cannot be stored for more than five years. This means that if the women have not used the embryos themselves within five years, they will be destroyed. Women can choose to allow the embryos to be used 'for science', but embryo donation is not possible in Denmark. Therefore, the women only have five years to decide what to do.

As a result of this five-year limit, women are habitually reminded of their decision to potentially destroy the frozen embryos. Women who have embryos cryopreserved at private clinics often get a letter every year asking them if they want to keep the embryo cryopreserved or to destroy it. Women who have embryos cryopreserved in public clinics do not get such a letter every year, but they are made well aware of the five-year limit. Mette, who was pregnant with twins at the time of the interview, said:

> Dagmar, the child I have now, she is two years old, and she was the second egg. The first egg didn't stick. And then they looked at my eggs, and those that were good, they asked, should we freeze them? At that moment, I thought, I am just having Dagmar, so whatever! Just freeze. Then I think one first realises when you get the letter, once a year, and the letter says, 'should we destruct your eggs, or should we continue to freeze them?' I think you get a feeling of ownership. I thought: 'hell no, you must not destruct my children!' [...] And I feel like that way of formulating it... I actually think that was what triggered it for me [to want to have another child]: The letter and the wording of destruction. I mean, I see it as my children or some kind of finished creatures, just on frost. (Interview with Mette 2018)

The eggs Mette mentions are the embryos from an IVF treatment, which she kins as children by fixing them in a genetic imaginary. But she also articulates an intensification of the feeling of time which some women experience when they have embryos 'on ice'. This intensification is brought to the fore by the letter sent from the clinic. Mette, as well as many other interviewees, do not like the use of the word 'destruction' in relation to the embryos. It seems to activate a feeling of responsibility towards the embryos, which pushes the women to want another pregnancy rather than destroying the embryos. The letter further reminds the women about the five-year time limit which seems to accelerate the feeling of time and the investment the embryos represent. Whereas at first ART and cryopreservation may feel like synchronising technologies that enable women to 'catch up on time', the five-year rule seems to intensify the need for not running out of it. This makes the women even more affectively connected to the embryos, thus further animating them.

Conclusion: Affective Animacy

It is in the context of these various circuits of animation that we are to understand the cryopreserved embryonic matter. The women in this study attempt to fix the meaning of liminal matter, to categorise it as either purely matter (as in the case of abortion) or as latent, valued kin (as in case of the cryopreserved embryos). The ontological status of the embryos seems to shift between matter, object, and subject, causing them to accumulate even more meaning and affect.

As mentioned above, affects gain value when circulated. The cryopreserved embryos are mediated through complex onto-epistemological circuits of changing meanings, ontologies, affective investments, and linguistic positions. The cryopreserved embryos are never fully fixed in meaning, but they appear and disappear in and vis-à-vis different visualising, temporal, and reproductive technologies. They can become regarded as agents in temporal processes and emerge in different affective arrangements as a consequence.

In a more materialist vein, I argue that the circulation and intensification the embryonic matter undergoes result in it becoming *affectively animated*. The ontological and emotional choreography, to use Thompson's and Adrian's concepts, or the normative, rhetorical, material, discursive and grammatical dance of the embryo—as both already sibling-becoming-child and 'mere cells'—makes the embryos objects of affective animacy. Embryos are potentialities; their liminal and latent status makes them both ontologically ambivalent and overdetermined. On the one hand, the embryos enable and perform temporal resynchronisation and lift their owners into other potential temporal orders by promising synchronisation to normative temporalities and life structures. On the other hand, the embryo haunts its owners with an ongoing intensification of time. This temporal movement and re-orchestration affectively embed the embryos in kinning relations.

Affective animacy constitutes both the emergence of the embryonic matter as an animated object, as the affects attached to the matter are what allows the embryos to emerge conceptually (Adrian, 2006; Ahmed, 2004). At the same time, the vibrant matter (Bennett, 2010) of the embryo forces and intensifies our affects towards it. Thus the embryonic matter represents onto-epistemological emergence in the very processes of animation.

Notes

1. The woman or couple must give approval for the national authority to gather information from doctors, authorities, etc. If she/they do not approve and the doctor cannot determine her/their suitability, she/they will be denied access to treatment. If they approve, the authorities will assess their suitability at least according to these concerns: (1) problems of addiction in the home, (2) the mental state of the woman/the couple, (3) matters that lead to placement of the child outside the home, and (4) whether either of the parents already have children placed outside the home. See https://www.regionsyddanmark.dk/wm519890 (in Danish).
2. For analyses and discussion of the ways in which the five-year limit emerged and is understood, see Kroløkke et al. (2020).

References

Adams, V., Murphy, M., & Clarke, A. (2009). Anticipation: Technoscience, life, affect, temporality. *Subjectivity, 28*(1), 246–265.

Adrian, S. (2006). *Nye skabelsesberetninger om æg, sæd og embryoner. Et etnografisk studie af skabelser på sædbanker og fertilitetsklinikker* [New narratives of creation of egg, sperm and embryos. An ethnographic study on sperm banks and fertility treatments]. PhD thesis, Linköping University.

Ahmed, S. (2004). *The cultural politics of emotion*. Edinburgh University Press.

Balsamo, A. (1995). *Technologies of the gendered body: Reading cyborg women.* Duke University Press.

Bennett, J. (2010). *Vibrant matter: A political ecology of things*. Duke University Press.

Berker, T., Hartmann, M., Punie, Y., & Ward, K. (2005). Introduction. In T. Berker, M. Hartmann, Y. Punie, & K. Ward (Eds.), *Domestication of media and technology* (pp. 2–17). Open University Press.

Casper, M. J. (1998). *The making of the unborn patient. A social anatomy of fetal surgery*. Rutgers University Press.

Chen, M. Y. (2012). *Animacies: Biopolitics, racial mattering and queer affect*. Duke University Press.

Cherry, J. L. (1992). *Animism in thought and language*. PhD thesis, University of California.

Cromer, R. (2018). Saving embryos in stem cell science and embryo adoption. *New Genetics and Society, 37*(4), 362–386.

Cromer, R. (2019). Racial politics of frozen embryo personhood in the US anti-abortion movement. *Transforming Anthropology, 27*(1), 22–36.

Eng, D. L. (2010). *The Feeling of Kinship: Queer Liberalism and the Racialization of Intimacy*. Duke University Press.

Franklin, S. (1999). Dead embryos: Feminism in suspension. In L. M. Morgan & M. W. Michaels (Eds.), *Fetal subjects, feminist positions* (pp. 1–82). University of Pennsylvania Press.

Freeman, E. (2010). *Time binds: Queer temporalities, queer histories*. Duke University Press.

Halberstam, J. (2005). *In a queer time & space – Transgender bodies, subcultural lives*. New York University Press.

Howell, S. (2006). *The Kinning of foreigners: Transnational adoption in a global perspective*. Berghahn Books.

Krøløkke, C. (2018). Frosties: Feminist cultural analysis of frozen cells and seeds documentaries. *European Journal of Cultural Studies, 22*(5–6), 528–544.
Krøløkke, C., & Nebeling Petersen, M. (2017). Keeping it in the family: Debating the bio-intimacy of uterine transplants and commercial surrogacy. In R. M. Shaw (Ed.), *Bioethics beyond altruism. Donating & transforming human biological materials* (pp. 189–214). Palgrave Macmillan.
Krøløkke, C., Myong, L., Adrian, S. W., & Tjørnhøj-Thomsen, T. (2016). Introduction. In C. Krøløkke, L. Myong, S. W. Adrian, & T. Tjørnhøj-Thomsen (Eds.), *Critical kinship studies*. Rowman and Littlefield International.
Krøløkke, C., Petersen, T. S., Herrmann, J. R., Bach, A. S., Adrian, S. W., Hansen, R. K., & Nebeling Petersen, M. (2020). *The cryopolitics of reproduction on ice: A new Scandinavian ice age*. Emerald Group Publishing.
Laclau, E., & Mouffe, C. (1985). *Hegemony and socialist strategy*. Verso.
Mohr, S., & Herrmann, J. R. (2021). The politics of Danish IVF: Reproducing the nation by making parents through selective reproductive technologies. *BioSocieties*. https://doi.org/10.1057/s41292-020-00217-1
Møller, K., & Nebeling Petersen, M. (2017). Bleeding boundaries: Domesticating gay hook-up apps. In R. Andreassen, M. Nebeling Petersen, T. Raun, & K. Harrison (Eds.), *Mediated intimacies: Connectivities, relationalities, proximities* (pp. 208–223). Routledge.
Payne, J. G. (2016). Mattering kinship: Inheritance, biology and egg donation, between genetics and epigenetics. In C. Krøløkke, L. Myong, S. W. Adrian, & T. Tjørnhøj-Thomsen (Eds.), *Critical kinship studies*. Rowman and Littlefield International.
Radin, J. (2013). Latent life: Concepts and practices of human tissue preservation in the international biological program. *Social Studies of Science, 43*(4), 484–508.
Roberts, E. F. S. (2007). Extra embryos: The ethics of cryopreservation in Ecuador and elsewhere. *American Ethnologist, 34*(1), 181–199.
Rose, N. (2007). *The politics of life itself: Biomedicine, power, and subjectivity in the twenty-first century*. Princeton University Press.
Silverstone, R. (1994). *Television and everyday life*. Routledge.
Squier, S. M. (2004). *Liminal lives: Imagining the human at the frontiers of biomedicine*. Duke University Press.
Thompson, C. (2005). *Making parents: The ontological choreography of reproductive technologies*. MIT Press.
Thompson, C. (2014). *Good science: The ethical choreography of stem cell research*. MIT Press.

Vora, K. (2015). *Life support: Biocapital and the new history of outsourced labor.* University of Minnesota Press.

Waldby, C. (2019). *The oocyte economy: The changing meaning of human eggs.* Duke University Press.

Williams, C., Kitzinger, J., & Henderson, L. (2003). Envisaging the embryo in stem cell research: Rhetorical strategies and media reporting of ethical debates. *Sociology of Health and Illness, 25*(7), 793–814.

Zerubavel, E. (1981). *Hidden rhythms: Schedules and calendars in social life.* University of Chicago Press.

3

The Affective Temporalities of Ovarian Tissue Freezing: Hopes, Fears, and the Folding of Embodied Time in Medical Fertility Preservation

Anna Sofie Bach

Introduction

> So I had an ovary removed (…) It gave a sort of comfort, so to speak, that; (it) granted some calmness so I could begin the treatment and a feeling that we had taken action on this matter and had insurance. And then I went through the (cancer) treatment. (Interview with Agnete)

Though much of the social scientific scholarship about fertility preservation technologies has centred on commercial egg banking and the emerging practices of so-called social freezing, where people in good health freeze their gametes for later use (e.g., Carroll & Kroløkke, 2018; van de Wiel, 2015; Waldby, 2019), fertility preservation is also frequently undertaken based on "medical indication." As in the case of Agnete, quoted above, who had been diagnosed with aggressive cancer, fertility preservation is increasingly used in situations where future fertility is compromised by

---

A. S. Bach (✉)
Department of Sociology, University of Copenhagen, Copenhagen, Denmark
e-mail: aba@soc.ku.dk

© The Author(s), under exclusive license to Springer Nature Singapore Pte Ltd. 2022
R. M. Shaw (ed.), *Reproductive Citizenship*, Health, Technology and Society,
https://doi.org/10.1007/978-981-16-9451-6_3

either disease or medical treatment, such as chemo- and radiation therapy. Here, fertility cryopreservation offers a remedy against future infertility, a well-known side effect of life-saving medical treatment.

While belonging to the realm of what Franklin (1997) termed "hope technology," referring in her early work to in vitro fertilization (IVF), fertility preservation also captures the processes of biomedicalisation (Clarke et al., 2010) and contemporary anticipatory logics (Adams et al., 2009) that orient doctors and patients to risk management, prediction, and prevention (Rose, 2007). As noted by Adams et al. (2009), anticipatory practices are an affective state and a specific "way of actively orienting oneself temporally" (p. 247). Inevitably, fertility preservation, which offers the prospect of continuing reproductive capacity, involves a future orientation shaped not only by biomedical intervention but more generally by what Radin and Kowal (2017) have called "cryo-optimism."

As I will explore in this chapter, the freezing and transplantation of ovarian tissue, an emergent yet largely overlooked technology within the social sciences (for exceptions, see Bach & Kroløkke, 2020; Kroløkke & Bach, 2020), cannot simply be conceptualised as future oriented. Rather, it constitutes multiple temporalities and involves a particular kind of affective volleying between dread and hope (Adams et al., 2009, p. 250). Moreover, advancing the social scientific theorising of "medical freezing," the chapter explores how fertility preservation constitutes "portals to the past" (Radin, 2013) and new ways of bringing together of older and younger body-selves (Landecker, 2010). Drawing on qualitative interviews with 42 Danish women, the chapter presents an analysis of the embodied experiences of ovarian tissue cryopreservation and transplantation. The analysis is situated within a theoretical framework that combines feminist scholarship on reproductive technologies (Franklin, 1997; Kroløkke et al., 2019; Waldby, 2019) with STS theorisation of cryobanks (Landecker, 2010; Lemke, 2019; Radin, 2013; Radin & Kowal, 2017) and affect theory (Ahmed, 2006, 2010; Berlant, 2011; Duggan & Munoz, 2009). Focusing on the ways in which cryopreservation (re)organises reproductive and embodied temporalities and how risk management is experienced and navigated on the level of the individual, I explore the affects produced and circulated as ovarian tissue is extracted for cryopreservation, stored, and transplanted, and how these processes shape embodied identities.

In contrast to the cryopreservation of eggs, ovarian tissue freezing preserves fertility in a more holistic way. That is, it holds the ability to restart the menstrual cycle, including hormone production and ovulation, and depending on the transplantation site can make non-assisted conception possible (Jensen et al., 2017). Emphasising this *technicity* (Waldby & Mitchell, 2006), the analysis pays specific attention to how the reversal of (premature) menopause is potentialised and experienced as a way to (re)synchronise the post-cancer body with the pre-cancer self.

In what follows, I describe the technology of ovarian tissue freezing and transplantation in more detail, situating it in the Danish context. This section is followed by an overall introduction to the theoretical framework, and a section where I account for the methods, data, and analytical strategies. The analysis is divided into three parts, focusing on three ways in which embodied temporalities are "folded" (Mansfield, 2017) as ovarian tissue is frozen, thawed, and transplanted.

Ovarian Tissue Freezing in Denmark

Denmark is among the pioneering countries that began preserving human ovarian tissue as early as the late 1990s following a series of successful animal studies, but before efficacy in humans had been proven (Rodriguez-Wallberg et al., 2016). The first child born following an ovarian tissue transplantation was in Belgium in 2004. Since then, more than 130 children have been born following ovarian tissue transplantation across the world, including at least 20 in Denmark (Gellert et al., 2018). Around 100 Danish women have had tissue transplanted, while more than 1300 women and girls have had ovarian tissue frozen (Bach et al., 2020). In 2019, the American Society for Reproductive Medicine stated that ovarian tissue freezing was no longer an experimental procedure (Penzias et al., 2019).

Ovarian tissue freezing relies on the discovery that, from birth, the ovaries contain around two million resting, immature egg cells. Though more than half of these disappear before puberty, until the mid-thirties an average ovary contains thousands of immature egg cells. In contrast to the freezing of eggs, ovarian tissue freezing requires no prior hormonal stimulation, and so can be performed from day to day. Additionally, it represents the only option for prepubescent children who cannot be stimulated for egg retrieval (Kristensen & Yding Andersen, 2018). Once removed from the body, the cortex of the ovary, where immature egg cells are located in adults, is cut into smaller pieces which are frozen individually, allowing for several rounds of transplantation. Thawed ovarian tissue is normally placed in the remaining ovary or in the abdominal wall (Jensen et al., 2017).

In Denmark, ovarian tissue freezing is offered free of charge within the public healthcare system. However, this is only if the patient is found to be eligible for the procedure. According to guidelines provided by Danish fertility specialists, the risk of losing one's fertility has to be significantly high (>50%) and the chances of survival must be good (Jensen et al., 2017). There is no fixed age limit on ovarian tissue freezing, but the procedure is only offered if the ovarian reserve is estimated to be "good" measured by either biological age (below 35) or through biomedical testing of Anti-Müller Hormone (AHM) that indicates follicle count (Jensen et al., 2017). If the follicle count is low, the chances of success are slim since many egg cells are lost during freezing and transplantation. Finally, freezing and transplantation practices correspond to the age regulations of the Danish legal framework that restricts all kinds of assisted reproduction to the age of 45 (Act no. 902, 2019). This also limits IVF, which is often needed following ovarian tissue transplantation, to the age of 45.

Theorising Ovarian Tissue Freezing and Affective Temporalities

The advancement of cryotechnology has attracted much scientific and scholarly attention. Not only have engineers and biomedical professionals been preoccupied with developing the relevant equipment and techniques, bioethicists (e.g., Harwood, 2009; Mertes & Pennings, 2011; Petersen, 2020), science and technology scholars (e.g., Landecker, 2010;

3 The Affective Temporalities of Ovarian Tissue Freezing: Hopes... 55

Lemke, 2019; Radin, 2013, 2017), and social science scholars (e.g., Hudson, 2020; Inhorn et al., 2018; Kroløkke et al., 2019; Martin, 2010; van de Wiel, 2015; Waldby, 2019) are exploring and debating the evolving and diverse practices of cryobanking. In particular, this growing body of literature has theorised how cooling technologies alter the meanings of life and death (Lemke, 2019; Radin & Kowal, 2017) and, not least, how cryotechniques produce plastic biologies as temporal change is "blocked" or "paused," and "natural cycles" of decay are prevented (Lemke, 2019, p. 4). Much attention has been given to the ways in which the ability to "suspend life" (Lemke, 2019) and keep vital processes in a state of "latency" (Radin, 2013, 2017; Radin & Kowal, 2015) potentialise cryopreserved matter. Not only can bodily bits and pieces be turned into new kinds of temporally mobile "spare parts" (Landecker, 2010) within systems of donation and self-donation (Waldby, 2019; Waldby & Mitchell, 2006), they also become marketable within the new bioeconomy, where reproductive cells in particular have become big business (Almeling, 2011; van de Wiel, 2020; Waldby, 2019).

In contrast to sperm, eggs, and embryos, but similarly to umbilical cord blood, ovarian tissue freezing has primarily been established within a "self-donation model" (Waldby & Mitchell, 2006). Like other types of organ donation, ovarian tissue cross-donation would require the use of immunosuppressive drugs. As discussed by Gosden (2008), cross-donation is not perceived as ethically reasonable within a biomedical understanding, since infertility is not a life-threatening condition. Consequently, ovarian tissue has not (yet) been commercialised in the same way as frozen eggs and sperm. However, the ability of frozen-thawed tissue to reverse menopause is increasingly imagined and potentialised as a new remedy against menopause, also beyond cancer treatment (e.g., Kristensen & Yding Andersen, 2018). As I discuss elsewhere (Kroløkke & Bach, 2020), fertility specialists currently debate whether ovarian tissue freezing should be an option for more women, perhaps all women, to decrease the risk of osteoporosis, heart disease, etc. that follows from entering menopause—a bodily state increasingly understood as "being at risk" and to expand due to the raising life expectancy.

As highlighted in feminist scholarship, cryopreservation involves the possibility of controlling and reorganising embodied, reproductive temporalities (Krøløkke et al., 2019; Landecker, 2010; van de Wiel, 2015). Notably, as emphasised by Martin (2010), the sociotechnical imaginary of postponement appears much more socially and ethically controversial when it comes to the emerging practices of "social freezing" than those of "medical freezing" (e.g., see Chap. 5 in this volume) though fertility preservation arguably always spurs from the anticipation of a potentially infertile future (see also Krøløkke et al., 2019; van de Wiel, 2015). Though most research shows that women who avail themselves of "social" egg freezing do so out of necessity; for example, because they do not have a partner, rather than a deliberate wish to "postpone" parenthood (e.g., Inhorn et al., 2018), "medical freezing" sits more easily within the narrative of "salvage technology" (Radin, 2017) that has surrounded cryotechniques since the early days. People who are already positioned as patients are, arguably, more readily understood as "naturally" in need of preventative intervention.

Radin's notion of "salvage technology" (2017) highlights the importance of exploring how fertility preservation, and ovarian tissue freezing in particular, are novel experiences that reconnect differently temporalised bodily elements (Landecker, 2010); the experience of bringing that which has been "saved" in the past into an embodied present to change the future. In her work on epigenetics, Mansfield (2017) develops the notion of "folded futurity" to capture how the new scientific understanding of environmental impacts on genetic expression links the past, present, and future in new ways. Epigenetics, Mansfield argues, not only produces the idea of biological plasticity, it also transforms the meaning of time and generational connectedness, as the genetic make-up of multiple, still-only-prospective generations are affected in the present. Multiple futures are "folded" into the present, where the risks posed on those yet-to-be-born must be handled. Simultaneously, the present is extended into the future in terms of its effects on the generations to come (Mansfield, 2017). Though Mansfield's interest is in foetal vulnerability, I find the non-linear understanding of reproductive futurity and her concepts of "folding" and "extending" useful as a way of approaching how the past, present, and future are reorganised by ovarian tissue cryopreservation and transplantation.

Returning to the concept of "hope technology" (Franklin, 1997) and extending the point that anticipation is an affective state (Adams et al., 2009) through the work of Sarah Ahmed (2006, 2010), I approach fertility preservation as an affective "orientation device" (2006, p. 3) to centre the intentionality of fertility preservation as a biomedical intervention. Orientation, Ahmed argues (2006, p. 2), is "a matter of how we reside in space" and our orientations are continuously shaped by how we are directed by and towards certain objects. In Ahmed's work, hope is closely connected to the cultural expectation of happiness, which she understands as an affective form of orientation. Happiness, Ahmed points out, is assumed to follow some life choices and not others, and functions as a promise "that directs you towards certain objects, as if they provide you with a necessary ingredient for the good life" (2010, p. 54). We assume that happiness follows proximity to certain things and persons, for example, children. Following decades of feminist scholarship demonstrating how childrearing has been socially constituted as the meaning of life, particularly for women (Thompson, 2005), Ahmed positions the family as what she calls a "happy object," while emphasising that the happy family also "circulates through objects" (2010, p. 45). Through this framework, it becomes possible to understand frozen ovarian tissue as a "happy object" which offers proximity, through the latent potentiality of the tissue, to happy family life.

However, hope and the promise of happiness also give us "a specific image of the future" (Ahmed, 2010, p. 29), and provide the emotional setting for disappointment. As formulated by Duggan and Munoz (2009, p. 279), and echoing Adams et al. (2009), in this sense, hope is a risk, to which the fear of failure is always attached. Inhorn et al. (2017) capture this dualism by characterising medical egg freezing as "Janus-faced." As they emphasise, medical freezing often takes place at the intersection of life and death. Here, egg freezing offers the promise of life in the shape of both future childbearing and survival. Yet, future motherhood can never be guaranteed, especially not in the context of a life-threatening diagnosis (Inhorn et al., 2017). In this sense, fertility preservation acts as a concrete reminder of all that can be lost, one's life included. Exploring the ambitiousness of optimism, Berlant (2011), another affect theorist, has also underlined how "optimism might not *feel* optimistic" (p. 2). Coining the notion of cruel optimism, Berlant demonstrates how the emotional investment in an idealised fantasy of "the good life"-to-come is often painful in the present. "A relation of cruel optimism exists," Berlant notes (2011, p. 1) "when something you desire is actually an obstacle to your flourishing." In the context of assisted reproduction, this resonates with what Franklin (1997), and, in the Danish context, Koch (1989) have shown with women who struggle to end fertility treatment after years of unsuccessful attempts. Despite both mental, physical, and financial suffering, they are willing to go beyond their well-being in the present and "exhaust all possibilities" (Franklin, 1997) to achieve their dream of the future.

Capturing the entanglement of hope and fear, and providing a rich framework for understanding the temporal reorganisation enabled by the "latent potentiality" of cold storage, these perspectives allow for the exploration of the complex embodied experiences and affective temporalities constituted as ovarian tissue is extracted for cryopreservation, stored, and transplanted. Before I turn to the analysis, I present the empirical material in more detail.

Methods and Data

This analysis is based on interviews with 42 Danish women, conducted between 2017 and 2019, as part of a multi-sided ethnographic study of ovarian tissue freezing. All the women had ovarian tissue frozen due to serious disease, most often a cancer diagnosis. The women were recruited in two ways: cancer organisations distributed the invitation on social media (11 recruited) and invitation letters were sent to all living patients who had ovarian tissue transplanted in Denmark before the summer of 2018, after permission was obtained from the Danish Patient Safety Authority (N=92, 32 recruited for interview). The interviews were semi-structured, and conducted face-to-face by the author (except for one, which took place using a video call). All but two of the interviews were recorded and transcribed verbatim. In an attempt to preserve some of the emotional intensity of the conversations that evaporate when talk is "frozen" in text (Knudsen & Stage, 2015), transcripts include indications of

3 The Affective Temporalities of Ovarian Tissue Freezing: Hopes... 59

emotional expressions such as laughter or crying. All names used in this essay are pseudonyms, and quotes have been translated into English.

The material cuts across a variety of experiences with ovarian tissue freezing and transplantation. The semi-structured design of the interviews allowed for the examination of different trajectories apparent in the diverse experiences of these women, and for a sensitivity to the difficult nature of many of the stories, which concern not only the hardships of cancer or other serious conditions, but also involuntary childlessness, and years of failed attempts to get pregnant (though several of the women had achieved pregnancies through the transplanted tissue). The sample includes some of the earliest cases of both ovarian tissue freezing and transplantation in Denmark from the early 2000s, as well as stories from women who had tissue cryopreserved or transplanted shortly before I met with them. Consequently, some had to recall events and situations that took place more than a decade earlier, while others were in the midst of emotional turbulence, waiting to find out if the tissue was working.

These different circumstances contribute to the affects that emerge in the accounts, and interviewees' temporal positions in relation to cryopreservation and transplantation shapes what Knudsen and Stage (2015, p. 9) call the "emotional recollection." However, in an interview study, data on affect is always already a subjective *account* of affects (Knudsen & Stage, 2015), organised in accordance with the affective cultures available for making sense of one's experiences, through which the interviewee attempts to render themselves and their situations meaningful to the listener (Riessman, 2008). Extending the methodological point that interview accounts are always situational accomplishments (Riessman, 2008), I approach affects as produced and (re)circulated in the interaction of the interview. Not only might the interviewee be affected by the telling of the story, I also see myself "turning" the interviewee towards the past by asking them to elaborate on earlier experiences, or towards the future by asking about their thoughts on what to do with potential surplus tissue.

My analytical strategy consisted of close readings of the interview material, from which I have identified situations where the technoscientific practices of ovarian tissue freezing and transplantation constitute "affective orientations." That is, the recounting of "events" (Berlant, 2011, p. 5), highly laden with emotional intensity (hope, fear, sadness, joy, etc.) in which past and future(s) are envisioned, animated and (re)constituted in relation to the present. It has been an abductive process of moving back and forth between existing conceptualisations of the entanglement of cryotechnology, affect and time, and my empirical data. It exposed that the freezing and transplantation of ovarian tissue, in the context of prospective death, (re)organise affective temporalities in complex ways.

Folded Futurities: Survival and (re)constitution of Reproductive Autonomy

This section centres the event of diagnosis and the initial discussion of fertility preservation options as moments of affective and temporal orientation. While the factual contours of these situations differed between the interviewed women, most of them emphasised how time was very much at stake in the context of diagnosis. Treatment often had to be initiated immediately, sometimes within days. If chemotherapy is the first step in the treatment plan, there is only a narrow window of time for fertility preservation to occur. One interviewee recalled how the specialist treating her cancer had only been willing to postpone her chemotherapy for a day or two, with the explanation, "What am I to do with your tissue if you are dead?" (Interview with Carol). In addition to manifesting the possibility of an infertile future in the present, the discussion about fertility preservation also illuminates the imminent danger of the disease.

Yet the discussion about fertility preservation also generates hope, bringing the vision of (having) a future into the present. One interviewee, Agnes, even said: "My mom, you know, said I should do it, because I had to survive. So, I had to get out, so I would, you know, like have a future." In this way, Agnes captured how children, even prospective ones, function as tokens of futurity (Edelman, 2004; Mansfield, 2017). Nina, who was referred to the fertility clinic by the oncologist to discuss her options, also recounted about the experience:

3 The Affective Temporalities of Ovarian Tissue Freezing: Hopes... 61

It was strange, because you sat there and "you also need to have x-rays of your lungs to check if there is anything in the lungs and you need blood samples for your liver" and that sort of thing "and then you need to go to the fertility clinic." I found that to be… strangely out of place because… well, okay, I am going to do that too. I thought I was going to die, right? But in a way, it was also life-affirming, because indirectly she told me that "we don't believe that you are going to die. So therefore, we want to give you a chance to have children later," right? There was a lot of optimism in this possibility because it means people in the system believe that you'll survive. You don't really believe that yourself for the first few days.

Nina's account shows how the offer to preserve her fertility effectively oriented her towards the future. Herself a healthcare professional, Nina recognised that fertility preservation is also a matter of resources, meaning that within the public healthcare system it would only be offered if her chances of survival were good. Thus, to many of the women, the promise of future fertility, regardless of the statistical chances of a reproductive outcome, fuelled a more fundamental optimism.

Nina, who already had children, was not certain whether or not she wanted more. Rather, as she explained, what she sought to preserve was her ability to decide. Similarly, reflecting on why she chose to undergo invasive surgery, despite not imagining herself having children, Ulla said: "How should I explain it? It just became really important to have a chance. That it wouldn't be the cancer deciding. That it would still be my own choice." As I wish to emphasise in the following, the narrative of reproductive choice, which resonated among the interviewed women, should also be seen as a specific, culturally embedded way of embodying the technoscientific creation of possibilities. Another interviewee, Eva, who was about to decline ovarian tissue freezing, since she found the evidence of its success poor, but who nevertheless went to fertility counselling, said:

I remember that we came out of that conversation we had with (the doctor) completely… I mean high, because it was such a cool… also because a lot of what he talked about concerned a life afterwards. And that was simply impossible to imagine when you had the whole thing in front of you, right? Well, okay, there is a chance that you will get out on the other side of this.

And there is a life, and you will have all your choices back. Because suddenly, you are reduced to a patient who just has to do as; I mean you just have to jump on the train, right?

Eva's recollection showed emphatically the hope that she experienced following her consultation, which took place almost a decade before I talked to her. As in Nina's case, the discussion about fertility preservation oriented Eva towards the life that she would have afterwards, which was revitalising in itself. Eva stressed several times that fertility preservation was as much about restoring her sense of having choices as being "…tied to a specific wish for another child." In the context of cancer, which Eva describes as a fast-moving train that she just had to jump onto, fertility preservation was the only thing many participants felt they were able to make decisions about and control. Though the cancer train is fast-moving, being the passenger is a passive subject position. Fertility preservation reinstated Eva's sense of autonomy.

The restoration of autonomy that reproductive choice brings is reinforced by the technicity of frozen ovarian tissue and its attendant imaginaries (Bach & Kroløkke, 2020). This hinges on the tissue's ability to restore hormonal production, and to the number of eggs contained in a small piece of tissue. Regarding the decision between having ovarian tissue or eggs cryopreserved, a choice sometimes available, Milena said:

> You are in a shock situation and there are so many huge things that you cannot deal with in your head. And then it's a way to, I guess, postpone some decisions at some level. I mean, I felt like by doing it this way; obviously there are no guarantees, but I felt like perhaps this gave me more of an insurance. And that… that I didn't have to… yes, that it was possible to postpone the decision about children. And about how many.

As the quote highlights, Milena's decision was informed by the knowledge that ovarian tissue allowed her to preserve more eggs than the single round of oocyte retrieval that was possible before her cancer treatment began. She felt like ovarian tissue freezing provided her with a better chance of having several children. In contrast to the generational temporality of "deep time" (Waldby, 2019), which connects the (unborn) child to past generations through genetic kinship, the imaginary of siblings is a kinship temporality of the future. Many interviewees stressed the importance of providing children (those already born as well as those yet-to-be-born) with siblings in order to secure good lives for them. While the wish for siblings can be understood as a way to synchronise family life with heteronormative ideals of the nuclear family, in the context of disease, as many of the women explained, siblings make a child less lonely if their parent has relapse and dies. In this sense, sibling futurity is a future orientation constituted as much by fear and disease as by hope and family norms.

While fertility preservation is overall positively connoted and provides recipients with hope about the future, it is important to stress the point, also made by Adams et al. (2009), that within preventive and anticipatory biomedicine, hopes and fears entangle. I elaborate on this aspect in the next section, turning to how frozen tissue extends the past into the present through the practices of transplantation and risk management.

Extended Pasts: Navigating Risk, Trauma, and Fear

In this section, I focus on the ways in which the past is folded or extended (Mansfield, 2017) into the present, as thawed tissue is returned to the body or in the contemplating of future pregnancies. The interviews demonstrate how post-cancer reproductive planning renews disease-related feelings of fear and evolves around the management of risk.

Some of the interviewed women talked about the risk of preserving cancerous cells in the frozen tissue, which is also a discussion within the biomedical community, though the procedure is considered safe for many types of cancer (Maschiangelo et al., 2018). In the interviews, I did not introduce the topic systematically and only discussed it if brought up by the interviewee, as I aimed not to produce unnecessary fears. The interviews show that this fear maps onto the organisation of the procedure, during which the ovary is normally removed before chemo- or radiation therapy commences. While a means to preserve as much reproductive potentiality as possible, the tissue is extracted from a cancerous body. As one interviewee expressed it, it was at a time where her body "felt full of cancer," though she rationally stated that she knew that her cancer had not spread in the pelvic area. For other women, the risk was more concrete due to their type of cancer, for example, blood cancers. Breast cancer survivors with BRCA genes, which increase the risk of ovarian cancers, were also informed that the tissue would need to be removed again after a couple of years, whether they had the children they wanted or not. Importantly, none of the women interviewed for this study had refrained from transplantation due to this fear. However, this demonstrates that the frozen tissue connects interviewees to their pre-treatment bodies, extending risk from the past into the present and future. These accounts also illuminate how the technicity of ovarian tissue freezing charges it with potentially more complex layers of meaning than egg freezing.

The past also entangled with the present and future in the accounts about returning to the physical space of the hospital and resubjecting the body to invasive treatment in the form of transplantation surgery and, in many cases, IVF treatment. Sandra said:

> I was sitting in the consultation room in the fertility clinic and looking out the window at the same building and the same view as when I was sitting one floor above and was told the exact opposite, that 'well you are ill, and you might not have children.' That was… a really, really weird experience.

Though in a new context of a potentially happy event, fertility treatment creates a temporal loop causing Sandra to revisit the traumatic experience of diagnosis.

Many interviewees also expressed concern about when to pursue pregnancies and the pros and cons of waiting. All interviewees were advised to wait for at least one and often two years after their final treatment before pursuing pregnancy. The women explained that not only does the body need to regain physical strength and break down the last bits of medication, but the risk of relapse is also highest within the first two years. Timing, however, was even more pertinent in the cases of oestrogen-sensitive breast cancer, in which case patients are placed on anti-oestrogen medication for up to ten years. Several of the women in this situation described their relief that it was possible to pause this course of medication. However, they also talked about the difficulties of negotiating between their reproductive desires and the risks involved in pausing anti-estrogen medication. Moreover, as several of the women explained, "the pause" only grants a short window of time to conceive, be pregnant, and breastfeed, which several expressed a strong desire to do. Reflecting on an upcoming appointment with the doctors to discuss her future options, Milena talked in great detail about managing not only the timing, but also the insecurities connected to her and her partner's desire to have another child:

> It has to be timed in a sort of way. And… it is also a sort of weighing because it's also about, you know, that we obviously do want to; that we want this (another child). But of course, it can't be at the expense of… of my getting sick again. But, I mean, no one would be able to say that; after all, everything is uncertain. So, in some sense, we just have to go along with how we feel about it and the dreams we have. (…) And they can, you know, advise about what they know, but without really, I mean… you never really get an answer (laughs). Or at least that's my experience. That you have to go home and sort of weigh for and against it. Because, for example, I asked about… you know, that I would have to go back on the medication, because it was also an issue in relation to breastfeeding… and they said that I could breastfeed for six months and then I had to go back on the medication (…). But then I asked how long I needed to be on the medication before I can stop again. And they can't really give an answer. "Well for… a while" (laughs), and that's a bit tough. I mean, it's difficult to deal with all these indefinite issues and it's not; I don't blame them because I know it's because they don't know. And when there is no evidence-based knowledge, they cannot advise you to do such and such. But after all, they do have a better sense about it than I do (laughs).

Illuminating the emotional trouble of navigating reproductive desires in a context of risk, this extract also highlights the uncertainties of anticipatory medicine, especially within the realm of treatments that still have an experimental element ("they don't know"). While Milena trusted the advice given by the doctors, the quote also shows how, for the most part, she had to rely on her own sense of direction ("go along with how we feel about it"). Interestingly, Milena expressed concern that she would be seen as "greedy" and willing to run too great a risk in order to have another child. This made her conflicted about her wish to provide her child with a sibling close in age. Her concern, which resonated with other interviews, illuminates how risk predictions are not only biomedically mediated but also socially configured and negotiated. The circulation of hope and fear involves not only the sick person and the doctors but their friends and family.

Clearly, ovarian tissue transplantation is entangled with the management of risk. In the next section, I elaborate on this point turning to the experiences of fertility restoration. Here I wish to highlight how ovarian tissue transplantation is not only a remedy against the health consequences of (premature) menopause, such as osteoporosis, but also how the restoration of the menstrual cycle impacts the reconfiguration of post-cancer selves.

Extended Past as (re)synchronisation with Former Body-Selves

Though the original aim of ovarian tissue freezing is reproductive, the joy of regaining one's menstrual cycle comes up repeatedly in the interview material. Though many of the women had menopausal symptoms treated medically, the interview material is filled with stories about the discomfort of menopause, such as hot flushes, causing a variety of socially awkward situations. The interviewees noted the strangeness of having these symptoms at a young age. Some explain, with laughter, that they have shared the experience with their mothers, or that they became experts at advising women much older than themselves about relief strategies. "Premature menopause" disturbs the normative scripts of ageing, and so the reversal of menopause can function as a way of normalising the body.

For several of the interviewees, menopause reversal was an important aspect of the procedure. Reflecting on her decision to preserve ovarian tissue over eggs, Solveig said:

3 The Affective Temporalities of Ovarian Tissue Freezing: Hopes... 67

> If I simply had eggs retrieved, then there was the whole thing about, well, the leftover ovary I still have, what if it never started functioning again? I mean, I would go into menopause then. Boom. And I was, you know, I was not really ready for that at the age of 34. I couldn't really relate to that. But all in all, I will return; It's the way I am going to get back to the closest to normal. Or be most like myself again.

As the extract shows, the ability of the tissue to re-establish the menstrual cycle stood out as a way to suture her post-cancer body with her pre-cancer self. While some of the interviewees never regained their cycles regularly or for longer periods of time, most of the women talked about transplantation as a form of bodily revitalisation. For instance, Signe, one of the interviewees who underwent transplantation for menopause relief, explained:

> Somehow, it was the feeling of moving back in, quietly and slowly. To a house which had not been... I mean, to a house where the water had been shut off and the heat turned off, because no one really lived there. So, it was like, you know, having the machinery up and running again, the cycle turned on and having... all those things, which began gradually. And I feel like I am still moving into my body, more and more. Yes. And accepting it more and more and that it now looks like this, which not all other bodies necessarily do. And it's not because it's wrong, it's just because it went through a lot.

To Signe, regaining her cycle felt like reclaiming a body that had been out of her control, similar to Eva's sentiment about regaining reproductive choices. As Signe's account also illuminates, this restoration narrative is connected to survivorship and overcoming the hardships of disease.

To many interviewees, regaining a hormonal cycle was an affective marker of having succeeded and overcome their disease and, in this sense, of being able to return to the life trajectory from before they fell ill. Elisabeth said about the experience:

> It was... (laughs) it sounds kind of crazy, but suddenly I was really happy about having my period, right. I mean... actually it was... it sounds a bit; but it was like becoming a woman again. And I really missed it. The feeling

that the body reacted in all kinds of ways, you know. I mean, technically I had been in a kind of menopause for those six-seven years, right. So, it was a really good feeling. I was really happy and proud. Not just because now I can do something (about having a child), but also because it just felt good. And now I also have to go to the store to buy sanitary pads again and well… yeah, it's hard to explain, but somehow the body just came back to life in a sort of way.

Elisabeth's account, like many of the stories about regaining menstrual cycles, is full of joy and pride. Like Elisabeth, many were surprised by their own emotional response to bleeding, which they normally did not enjoy. As Elisabeth's quote reveals, the return of the menstrual blood nevertheless reconfigured her body as "normal" ("I also have to go to the store to buy sanitary pads") and, importantly, it restored her gender identity. Like others, Elisabeth spoke about how cancer treatment compromised not only her fertility, but also her femininity. While her hair had grown out and plastic surgery restored her breast, being resynchronised to the rhythmic temporality of the menstrual cycle fully restored her sense of womanhood. A sign that the tissue had revascularised, bleeding also constitutes an affective, as well as cultural, marker of proximity to childbearing ("Now I can do something").

Though in the end Elisabeth had to use egg donation, the transplanted tissue restored her sense of agency. As many of the women emphasise, the frozen tissue grants them the chance to pursue genetic reproduction and exhaust all their options (Franklin, 1997), which they are grateful for even if they do not succeed. Several of the women did not want to leave any tissue unused in the cryotank. Some wanted to stay out of menopause as long as possible; for others, it was also about reassembling a body taken apart by disease. While she had given up on having children, Marianne talked about having the rest of her tissue transplanted because she "had like a sense that I needed to have my body unified again. That it needed to be whole in some way." Echoing the other accounts, Marianne's account reflects how tissue transplantation is not only about fertility; it contributes to a more fundamental restoration of the post-disease body-self.

Concluding Reflections on the Affective Temporalities of Ovarian Tissue Freezing

This chapter examines how affective temporalities are constituted as ovarian tissue is frozen, stored, and transplanted in the context of serious disease and medical treatment. Being the first qualitative study focusing not only on decision-making, but also on the transplantation of ovarian tissue, the analysis casts light on the specific ways in which hope, fear, and joy entangle in medical freezing, which is crucial for theorising how it differs from egg and embryo freezing.

Throughout this analysis, I have highlighted how ovarian tissue freezing cannot be understood simply as a future-oriented "hope-technology." Following Ahmed (2014), the frozen-thawed tissue is "sticky" with both hope and fear. As emphasised, tissue transplantation involves a particular kind of risk management; procreative desires need to be navigated in accordance with the ongoing prediction of risk, continuously held against the chance of reproductive success. It is not simply the act of cryopreserving reproductive matter that should be understood as an anticipatory and preventative practice. Within the regime of medical freezing, speculative forecasting (Adams et al., 2009) continues to shape *when* ovarian tissue is transplanted, *if* it needs to be removed again (in the case of BRCA genes), and *how* the women, in collaboration with their doctors and partners, plan for pregnancies, not least the spacing between children. Risks and fears, reconnecting the otherwise cured patient to their diseased past, are co-constitutive for the ways in which the interviewed women navigate the options provided by cryopreservation.

Similar to the findings of Inhorn et al. (2017) and Hoeg et al. (2016), I have shown how ovarian tissue freezing comprises "a hope in the dark," although discussions about fertility simultaneously underline the precarity of the future. Emphasised by techno-scientifically constituted possibility, multiple futures continuously fold into the present (Mansfield, 2017). Like with egg freezing, the chance to pursue genetic parenthood restores a feeling of choice and reproductive autonomy. Adding to the feminist scholarship on fertility preservation, the chapter stresses how ovarian tissue transplantation involves a particular temporal reorganisation of the body, which is underexplored from an embodied perspective. While Landecker (2010) notes the emergence of "age chimaeras" (p. 216), I have sought to demonstrate how bringing reproductive matter from the past into the present is experienced as *(re)synchronisation* rather than transformation. The stored tissue comes to operate like "spare parts" (Landecker, 2010; Waldby & Mitchell, 2006), used to *repair* a body broken by medical treatment. Besides from enabling genetic kinship, regaining the menstrual cycle restores a sense of normality and properly timed ageing.

Yet, while not the primary aim of ovarian tissue freezing as it currently takes place in Denmark, the ability of ovarian tissue to reverse or postpone menopause holds perspectives beyond reproduction as also evidenced in the accounts about the revitalising sensation of regaining one's menstrual cycle. Whereas fertility preservation options alter our understanding of infertility treatment, the ability to reverse menopause further reshapes the idea of reproductive ageing. It fundamentally reconfigures the linear ontology of fertility from having a starting point (puberty) and one end point (menopause) in ways that need more exploration. This chapter has revolved around ovarian tissue freezing as it takes place in the context of serious disease, but the future of ovarian tissue freezing also lies beyond cancer treatment.

Acknowledgements Thank you to all the interviewed women who generously shared their experiences with ovarian tissue freezing and transplantation. Thank you also to the colleagues who commented on early drafts and to the editor of this volume and the reviewers for their constructive feedback.

References

Act NO. 902, 2019: Act on assisted reproduction. Retrieved November 23, 2020, from https://www.retsinformation.dk/eli/lta/2019/902

Adams, V., Murphy, M., & Clarke, A. (2009). Anticipation: Technoscience, life, affect, temporality. *Subjectivity, 28,* 246–265.

Ahmed, S. (2006). *Queer phenomenology. Orientations, objects, others.* Duke University Press.

Ahmed, S. (2010). *The promise of happiness*. Duke University Press.
Ahmed, S. (2014). *The Cultural Politics of Emotion*. Edingburgh University Press.
Almeling, R. (2011). *Sex cells: The medical market for eggs and sperm*. University of California Press.
Bach, A. S., & Kroløkke, C. (2020). Hope and happy futurity in the Cryotank: Biomedical imaginaries of ovarian tissue freezing. *Science as Culture, 29*(3), 425–449.
Bach, A. S., Macklon, K. T., & Kristensen, S. G. F. (2020). Futures and fears in the freezer: Danish women's experiences with ovarian tissue cryopreservation and transplantation. *Reproductive Biomedicine Online, 41*(3), 555–565.
Berlant, L. (2011). *Cruel optimism*. Duke University Press.
Carroll, K., & Kroløkke, C. (2018). Freezing for love: Enacting 'Responsible' reproductive citizenship through egg freezing. *Culture, Health and Sexuality, 20*(9), 992–1005.
Clarke, A. E., Shim, J. K., Mamo, L., Fosket, J. R., & Fishman, J. R. (2010). Introduction. In A. E. Clarke, L. Mamo, J. R. Fosket, J. R. Fishman, & J. K. Shim (Eds.), *Biomedicalisation. Technoscience, health and illness in the U.S* (pp. 1–44). Duke University Press.
Duggan, L., & Munoz, J. E. (2009). Hope and hopelessness. A dialogue. *Women and Performance: A Journal of Feminist Theory, 19*(2), 275–283.
Edelman, L. (2004). *No future: Queer theory and the death drive*. Duke University Press.
Franklin, S. (1997). *Embodied progress. A cultural account of assisted conception*. Routledge.
Gellert, S. E., Pors, S. E., Kristensen, S. G., Bay-Bjørn, A. M., Ernst, E., & Yding Andersen, C. (2018). Transplantation of frozen-thawed ovarian tissue: An update on worldwide activity published in peer-reviewed papers and on the Danish Cohort. *Journal of Assisted Reproduction and Genetics, 35*(4), 561–570.
Gosden, R. G. (2008). Ovary and uterus transplantation. *Reproduction, 136*, 671–680.
Harwood, K. (2009). Egg freezing: A breakthrough for reproductive autonomy? *Bioethics, 23*(1), 39–46.
Hoeg, D., Schmidt, L., & Macklon, K. T. (2016). Young female cancer patients' experiences with fertility counselling and fertility preservation – A qualitative small-scale study within the Danish health care setting. *Upsala Journal of Medical Sciences, 121*, 283–288.

Hudson, N. (2020). Egg donation imaginaries: Embodiment, ethics and future family formation. *Sociology, 54*(2), 346–362.

Inhorn, M. C., Birenbaum-Carmeli, D., & Patricio, P. (2017). Medical egg freezing and cancer patients' hopes: Fertility preservation at the intersection of life and death. *Social Science and Medicine, 195*, 25–33.

Inhorn, M. C., Birenbaum-Carmeli, D., Westphal, L. M., Doyle, J., Gleicher, N., Meirow, D., Dirnfeld, M., Seidman, D., Kahane, A., & Patrizio, P. (2018). Ten pathways to elective egg freezing: A binational analysis. *Journal of Assisted Reproduction and Genetics, 35*, 2003–2011.

Jensen, A. K., Kristensen, S. G., Macklon, K. T., Jeppesen, J. V., Fedder, J., Ernst, E., & Andersen, C. Y. (2017). Outcomes of transplantations of cryopreserved. *Human Reproduction, 30*(12), 2838–2845.

Knudsen, B. T., & Stage, C. (2015). Introduction: Affective methodologies. In B. T. Knudsen, C. Stage, & C. (Eds.), *Affective methodologies: Developing cultural research strategies for the study of affect* (pp. 1–22). Palgrave Macmillan.

Koch, L. (1989). *Ønskebørn: Kvinder og Reagensglasbefrugtning (Wished-for Children: Women and IVF)*. Rosinante.

Kristensen, S. G., & Yding Andersen, C. (2018). Cryopreservation of ovarian tissue: Opportunities beyond fertility preservation and a positive view into the future. *Frontiers in Endocrinology, 9*, 347.

Kroløkke, C., & Bach, A. S. (2020). Putting menopause on ice: The cryomedicalization of reproductive aging. *New Genetics and Society, 39*(3), 288–305.

Kroløkke, C., Petersen, T. S., Herrmann, J. R., Bach, A. S., Adrian, S. A., Klingenberg, R., & Nebeling Pedersen, M. (2019). *The cryopolitics of reproduction on ice. A new scandinavian ice age*. Emerald Publishing Limited.

Landecker, H. (2010). Living differently in time: Plasticity, temporality and cellular biotechnologies. In J. Edwards, P. Harvey, & P. Wade (Eds.), *Technologized images, technologized bodies* (pp. 211–236). Berghahn Books.

Lemke, T. (2019). Beyond life and death. Investigating cryopreservation practices in contemporary societies. *Soziologie, 48*(4), 450–466.

Mansfield, B. (2017). Folded futurity: Epigenetic plasticity, temporality, and new thresholds of fetal life. *Science as Culture, 26*(3), 355–379.

Martin, L. J. (2010). Anticipating infertility: Egg freezing, genetic preservation, and risk. *Gender and Society, 24*(4), 526–545.

Masciangelo, R., Bosisio, C., Donnez, J., Amorim, C. A., & Dolmans, M. M. (2018). Safety of ovarian tissue transplantation in patients with borderline ovarian tumors. *Human Reproduction, 33*, 212–219.

Mertes, H., & Pennings, G. (2011). Social egg freezing: For better, not for worse. *Reproductive Biomedicine Online, 23*, 824–829.
Penzias, A., Bendikson, K., Falcone, T., Gitlin, S., Gracia, C., Hansen, K., Hill, M., Hurd, W., Jindal, S., Kalra, S., Mersereau, J., Odem, R., Racowsky, C., Rebar, R., Reindollar, R., Rosen, M., Sandlow, J., Schlegel, P., Steiner, A., & Tanrikut, C. (2019). Practice committee of the American society for reproductive medicine. Fertility preservation in patients undergoing gonadotoxic therapy or gonadectomy: A committee opinion. *Fertility and Sterility, 112*, 1022–1103.
Petersen, T. S. (2020). Arguments on thin ice: On non-medical egg freezing and individualisation arguments. *Journal of Medical Ethics, 47*(3), 164–168.
Radin, J. (2013). Latent life: Concepts and practices of human tissue preservation in the inter-national biological program. *Social Studies of Science, 43*(4), 484–508.
Radin, J. (2017). *Life on ice: A history of new uses for cold blood.* Chicago University Press.
Radin, J., & Kowal, E. (2017). Introduction: The politics of low temperature. In J. Radin & E. Kowal (Eds.), *Cryopolitics: Frozen life in a melting world* (pp. 3–25). The MIT Press.
Riessman, C. K. (2008). *Narrative methods for the human sciences.* Sage Publications Inc.
Rodriguez-Wallberg, K. A., Tanbo, T., Tinkanen, H., Thurin-Kjellberg, A., Nedstrand, E., Kitlinski, M. L., ... Andersen, C. Y. (2016). Ovarian tissue cryopreservation and transplantation among alternatives for fertility preservation in the Nordic countries – Compilation of 20 years of multicenter experience. *Acta Obstetericia and Gynecologia Scandinavia, 95*(9), 1015–1026.
Rose, N. (2007). *The politics of life itself: Biomedicine, power, and subjectivity in the twenty-first century.* Princeton University Press.
Thompson, C. (2005). *Making parents. The ontological choreography of reproductive technologies.* The MIT Press.
van de Wiel, L. (2015). Frozen in anticipation: Eggs for later. *Women's Studies International Forum, 53*, 119–128.
van de Wiel, L. (2020). The speculative turn in IVF: Egg freezing and the financialization of fertility. *New Genetics and Society, 39*(3), 306–326.
Waldby, C. (2019). *The oocyte economy. The changing meaning of human eggs.* Duke University Press.
Waldby, C., & Mitchell, R. (2006). *Tissue economies: Blood, organs, and cell lines in late capitalism.* Duke University Press.

4

Trans Narratives of Fertility Preservation: Constructing Experiential Expertise Through YouTube Vlogs

Alex Ker and Rhonda M. Shaw

Introduction

Fertility preservation, for medical or elective reasons, is the process of obtaining and freezing gametes (ovarian eggs or sperm) or reproductive tissue to retain a person's future fertility options for family-building.[1] Fertility preservation and reproductive rights for transgender (trans) people are gaining increasing attention in both the medical profession and among trans people, whose gender is different from their sex assigned at birth.[2] While trans people's decision to medically transition[3] historically meant ruling out the possibility of having genetically related children, recent developments in assisted reproductive technologies, and research examining the impacts of gender-affirming hormone therapy on fertility, are enabling more trans people to plan their reproductive futures with the possibility of biologically related children in mind.

A. Ker (✉) • R. M. Shaw
School of Social and Cultural Studies, Te Herenga Waka—Victoria University of Wellington, Wellington, New Zealand
e-mail: alex.ker@vuw.ac.nz; rhonda.shaw@vuw.ac.nz

Despite such progress, many trans people's reproductive choices are still limited by misinformation, or lack of information, on fertility preservation and transitioning which they receive from healthcare providers. Research shows that a significant number of trans people are still not fully informed by healthcare professionals about options to preserve their fertility before starting gender-affirming hormone therapy. In a large non-clinical study of trans fertility preservation recently undertaken in Australia, Riggs and Bartholomaeus (2018) found that while 95% of survey participants believed the option to preserve fertility should be offered to trans people, only 23% of respondents reported being given advice or counselling about fertility preservation options from a healthcare professional. The current lack of discussion around fertility preservation may reflect healthcare professionals' wider lack of competence or confidence around providing gender-affirming healthcare (Vance et al., 2015), or stigma towards trans people in healthcare settings more broadly (Ellis et al., 2015; Poteat et al., 2013; Stroumsa et al., 2019).

Research on trans people's interactions with fertility services indicates the prevalence of cisnormativity in healthcare settings (Epstein, 2018). The assumption that everyone is cisgender, that their gender aligns with the sex they were assigned at birth, is evident in some fertility service providers' failure to acknowledge a trans person's self-determined gender as different from their sex assigned at birth (James-Abra et al., 2015), and the conflation of reproductive capacities (whether someone has sperm or ovarian eggs) with gender-specific parenting roles (being a mother or father) (James-Abra et al., 2015; Riggs, 2019; Riggs & Bartholomaeus, 2018). Consequently, trans people's unique needs may be overlooked when accessing fertility preservation information or services, such as the psychological effects of delaying transitioning or being misgendered by healthcare staff. As Pearce (2018) notes, although cisnormativity does not always manifest in overt acts of transphobia, it nonetheless creates barriers for trans people receiving necessary healthcare.

It is also worth noting that while some trans people desire to nurture children, some may not want to parent. For instance, they might already have children and not want more. Alternatively, some individuals may not want to be genetically related parents but could consider other pathways to parenthood such as adoption, fostering, or guardianship (Chiniara et al., 2019; Riggs & Bartholomaeus, 2018). So, while the right to engage in family-building projects is commonly regarded as central to full participatory citizenship (Turner, 2008), there is a risk that biomedical guidelines offering healthcare advice and consumer information around fertility preservation could reinforce the importance of biological parenthood as an integral part of the adult life-course. This could be detrimental to the wellbeing of individuals who do not aspire to heterosexual relational norms and do not want to have children. As several authors argue, encouraging trans people to consider fertility preservation as an insurance strategy 'just in case' they change their minds about future family creation not only perpetuates a genetic norm about biological relatedness as ideal (Riggs, 2019), but produces trans individuals as 'patient consumers' of reproductive services who then become 'parents-in-waiting' (Mamo, 2013).[4]

Existing research gives insight into trans people's views and experiences on fertility preservation as a family-building strategy in a clinical context, but there has been little to no qualitative research to date on how fertility preservation is discussed among trans people outside of healthcare settings. Trans communities, like other groups of people with shared experiences of embodiment, play a vital role in sharing information about experiences which are otherwise absent from dominant sources of healthcare information (Pearce, 2018). Over the past decade, the internet has facilitated connections between trans people across the world, who are able to share stories of accessing gender-affirming healthcare beyond spatial and temporal boundaries. However, while our observations suggest that several fertility clinic websites and blogs feature cisgender people's fertility preservation stories, trans-specific information or representation on these online sources remain comparatively scarce (see also Wu et al., 2017). New Zealand[5] fertility clinic websites, for example, have historically privileged cisgender experiences. Until very recently, only one New Zealand clinic, Repromed, made mention of gender diversity on their website. In 2020, the largest New Zealand clinic, Fertility Associates, included a section on LGBTQ+ people which contains a brief reference to 'Transgender Fertility'.

As a tool to provide the information neglected by dominant healthcare providers, the public video-sharing platform YouTube has become popular among trans people for sharing experiences, building community, and finding support (Eckstein, 2018; Horak, 2014; Miller, 2017; O'Neill, 2014; Raun, 2010, 2015a, 2015b). Through filming and uploading video blogs (vlogs), trans people can both archive their own transitions for personal benefit (Raun, 2015b), and use videos as educational tools to teach trans people about trans-related issues (Miller, 2017). The videos' audio-visual format means that vlogs offer viewers intimate ways of understanding how being trans and transitioning may feel, sound, and look like. The interactive nature of YouTube further enables trans people to exchange information and experiences with viewers and other YouTubers through comments, messages, or video responses.

YouTube, like other public online information sources, has significantly increased lay people's access to medical knowledge that was previously thought of as exclusive to the medical community (Dewey, 2008). In this sense, YouTube may act as a way for trans people to fill the 'gaps' in medical education and resources about accessing healthcare. Most existing literature on trans vlogs has focused on trans people's presentation of their gender and transition. Considering the increasing awareness of and access to gender-affirming healthcare, it is equally important to consider the epistemic value of YouTube in relation to trans people's experiences of fertility and the prospect of family-building. Further, the current lack of discussion between healthcare professionals and trans people about their reproductive futures (Riggs & Bartholomaeus, 2018) indicates the need to better understand how trans people themselves are producing and sharing knowledge about transition-related healthcare outside of clinical settings.

Although trans people have been documenting various aspects of transitioning on YouTube since 2006 (Raun, 2018), the first vlog on trans fertility our search located was uploaded to YouTube in 2014. Since then, the growing number of trans people creating vlogs about fertility preservation reflects a discursive shift from trans fertility and parenthood being perceived as a stigmatised issue—largely due to the misperception that trans people might make 'bad parents' because of their gender (Chen et al., 2018; Murphy, 2012)—to an important subject for many trans people. This chapter sets out to understand the functions of trans people's YouTube videos on fertility preservation, and the types of knowledge vloggers and viewers produce and exchange through vlogging. To discuss the YouTube vloggers' accounts of their experiences and decision-making, we draw on Pearce's (2018) definition of an epistemic community as 'a collection of individuals and groups among whom complementary forms of knowledge and expertise circulate' (p. 164). As such, we use the term experiential expert (Akrich et al., 2008; Wehling et al., 2015) in our chapter to refer to vloggers as people with intimate knowledge and experience of their bodies and identities in relation to the healthcare issues they discuss. We apply the term experiential expert rather than 'lay expert', as the vloggers do not necessarily contest or seek to add to medical knowledge, but attribute value to the production of 'alternative ways of understanding and articulating their experience' (Akrich et al., 2008, 17).

Methodology

The present study draws on a contextualist framework to understand trans people's fertility preservation experiences, in reference to trans people's knowledge production online. Contextualism is an epistemology which proposes that knowledge is contingent on and reflective of the situations in which it is produced (Madill et al., 2000). Within a contextualist framework, researchers consider factors which affect participants' understandings of phenomena, researchers' positionality, cultural meaning systems in which experiences are located, and the validity of interpretations (Pidgeon and Henwood, cited in Madill et al., 2000, 9). In contrast to realist perspectives, which tend to be critical of researcher bias, Madill et al. (2000) contend that 'the empathy provided by a shared humanity and common cultural understanding [within a contextualist framework] can be an important bridge between researcher and participant and a valuable analytic resource' (p. 10).

The use of a contextualist framework is relevant to the study this chapter is based on. The vlog study is part of a larger project on reproductive futures, which was designed by Author 2, a cisgender sociologist, with research experience in the field of assisted human reproduction. To assist with the empirical research discussed in this chapter, Author 1 adopted a trans epistemological standpoint, interpreting the vlogs (and vloggers' experiences) as a person for whom these vlogs are intended. When watching vlogs, the authors took care to avoid abstracting vloggers' stories or using them as objects of study (Radi, 2019). To do this, Author 1 became familiar with vloggers' YouTube channels and their other public social media sites (if applicable) to understand vloggers' interests and motivations for creating vlogs, and the degree to which they were comfortable sharing their lived experiences online.

Data Collection and Analysis

Initial videos were found through keyword searches on YouTube, using a combination of terms including transgender, fertility, egg freezing/harvesting, freezing eggs/sperm, banking/preserving, story, experience, and cryopreservation. Author 1 subsequently found videos through YouTube's recommendations sidebar. A total of 51 videos were collected over a 6-month period between November 2019 and April 2020. Of these, 38 videos met the inclusion criteria (see Appendix) and were included in the final data for analysis. Vloggers collectively represented various ethnicities, gender identities and expressions, ages, sexualities, and countries of residence including New Zealand, Australia, Canada, India, Germany, USA, and the United Kingdom. A total of 362 minutes of footage were analysed, with the average length of videos being 6.9 minutes. Data consisted of vlogs made by both non-professional and 'micro-celebrity' vloggers (Raun, 2018), some of whom generated revenue off their YouTube channel through advertisements and large numbers of subscribers.

In the current study, the authors undertook a conventional content analysis (Hsieh & Shannon, 2005) to describe patterns and differences across vloggers' experiences of fertility preservation, and the types of information vloggers share. The videos were each watched several times, noting the content of vloggers' talk, the nature of their YouTube channels, and any relevant comments from viewers responding to the videos. Using this data collection process, two main types of vlogs were identified. In the first category, vloggers documented the gamete (sperm or ovarian egg) storage process they had gone through, or were going through at the time of filming. A few vloggers documented their experiences and changes in their desires through a series of videos over time. The second category consisted of talking-head-style videos in which vloggers discussed the reasons they chose (or did not choose) to preserve their fertility. These videos tended to be more about the reasons behind vloggers' decision-making, rather than describing the gamete storage process itself. Author 1 transcribed excerpts of the videos verbatim and annotated relevant visual cues. Codes were subsequently identified and developed from the annotations, and grouped accordingly to construct a description of the videos' content.

Ethical Considerations

Scholars have previously discussed the tensions around using data collected from publicly available social media platforms (Markham, 2012; Patterson, 2018). Some scholars have noted that using publicly available YouTube videos as data does not require research ethics clearance because YouTube copyright agreements enable the public use of videos (see Giles, 2017). However, even if vloggers choose to make their YouTube videos publicly available, this does not imply that vloggers intend for, or agree to, their videos to be used for research purposes. As such, using YouTube content without creators' informed consent, even when data is anonymised may be construed as covert observation and thus pose ethical risks around breach of privacy and psychological harm to the vloggers.

Considering these tensions, the authors took several steps to ensure transparency and minimise the risk of harm. For example, we noticed six videos which were publicly available at the time of data collection and analysis had subsequently been marked as 'private', and therefore no longer available to the public. We amended our findings to ensure that we did not make any direct reference to vloggers who made their videos private. Although we cannot know vloggers' reasons for their decisions, it is a reminder of the dynamic nature of digital media platforms such as YouTube. In reflecting on working with YouTube videos as data, Patterson (2018) encourages researchers to be aware of the 'cultural norms of the [online] space […] with acknowledgement that these norms are not stagnant but are subject to continual change' (p. 765). Content creators on YouTube may decide at any point to use YouTube for personal documentation, rather than to share their experiences with others, and we responded to these changes accordingly throughout the research process, keeping in mind that intimate moments disclosed on social media platforms are not necessarily 'shared for all to see' (Garde-Hansen and Gorton, cited in Nebeling Petersen et al., 2018, 1).

Sensitivity to data should also be prioritised when research topics include potentially intimate discussions around bodies and transitioning. As vloggers do not always provide contact details on their YouTube channels, it can be difficult to contact them to ask for their permission. To ensure informed consent, we contacted vloggers who provided contact details on their YouTube channels to explain the study and ask for their permission to use their quotes if relevant. Of the 15 vloggers contacted, six replied and gave their written consent. The present study was approved in 2020 by the Victoria University of Wellington Human Ethics Committee (reference number 0000028281).

Findings

Four main aspects of vlogs were identified throughout our analysis, pertaining to the types of information and knowledge shared through vloggers' videos. First, vloggers positioned themselves as experts of their own experiences, rather than critiquing or claiming to have medical knowledge about fertility preservation. They discussed the uncertainties and tensions between transitioning and planning their reproductive futures. Vloggers also spoke about and visually represented their bodies and emotions in intimate ways on screen, which offered viewers authentic insight into fertility preservation processes. Finally, many vloggers explicitly negotiated the cisnormative knowledge in existing reproduction discourses and healthcare settings.

Vloggers as Experiential Experts

Throughout their videos, vloggers frequently discussed the lack of trans-specific information as the reason why they decided to document their own journeys of fertility preservation. This motivation to share their own experiences for educational or informational purposes suggests the integral role of YouTube videos in creating and sharing trans-specific knowledge. Here, vloggers intended to fill a gap in trans-specific information which they perceived to be lacking in online content on fertility preservation, as the following comments illustrate:

> I've found absolutely no material filmed or—a little bit written by other trans guys [...] so be the change you want to see in the world, I guess? (FT, 2016)

> [Trans fertility preservation] isn't usually talked about, which is why I'm gonna touch on it now. Plus, when I was searching, I couldn't find a-a-any information. (CC, 2018)

By proceeding to share their experiences or 'stories', as some vloggers framed it, they intentionally shared particular knowledge grounded in their lived experience. Vloggers achieved this by positioning themselves as storytellers or experts over their own identities, thoughts, and perspectives, rather than claiming medical authority over the topic:

> Just a disclaimer—I'm not a professional egg freezer, I don't know my stuff on this as well as, maybe, medical professionals. I'm not claiming to know everything, I'm just try'na put across what happened to me, my experience with it, and maybe it will help out you guys. (NF, 2018)

> I'm not gonna be the one saying, 'you to need to take this milligram, this gram, this pill, this shot', I am not that kind of bitch, I do not feel comfortable doing that. (GG, 2018)

> I'm not the expert, I haven't done a lot of research into it. (JE, 2018)

In setting out the limits of their experiential knowledge, vloggers positioned themselves as lay people who could speak to their own experiences of fertility preservation, thereby helping viewers relate to their lived experiences. Interestingly, the construction of experiential expertise in vlogs contrasts with the positioning of lay people as 'patient-experts' or 'patient-activists' in similar lay epistemic communities (Akrich, 2010; Pearce, 2018; Tattersall, 2002), whose members advocate for themselves by drawing on medical and scientific discourses to challenge medical authority (Dewey, 2008).

Despite not being experts in medical aspects of fertility preservation, vloggers demonstrated their experiential expertise by giving advice to viewers, who were typically assumed to be trans people considering fertility preservation. Vloggers integrated pieces of advice throughout their narratives about fertility preservation, based on what they had learnt from their own experiences. For example:

> I hope that helps anybody who's contemplating it… Personally, if I didn't try and me and my girlfriend wanted kids in the future, and I didn't try this and I thought 'stuff it, I want the hormones now', I know I'd be the person to regret it in the future and I don't want people to go through that as well, because obviously, although you're thinking about what's best for you now, you also need to think about what's best for you in the future. (OA, 2015)

> If you are thinking about starting oestrogen, I would strongly urge you to consider the option of sperm banking, of storing your sperm, if you can afford it, ah, not necessarily because you want to use it in the future, but just so you have the option if you do change your mind in the future. (TTM, 2019)

This advice, which is marked temporally by past and future considerations and encouraging viewers to store gametes 'just in case' (Riggs, 2019), also serves to affirm trans people's bodily autonomy and personal choice. As one vlogger, JE (2018), shared, 'this is not your partner's, this is not your parent's, this is not your friend's decision. This is not your doctor's decision to make, this is yours'. The vlogs' advice-giving function constructs experiential knowledge as something viewers can use to inform their decision-making, and to not make the same 'mistakes' others did in the past due to the lack of information around fertility preservation options for trans people.

Some vloggers encouraged viewers to engage in discussions about the content of their videos, through commenting or asking questions. YouTube comment threads provide a virtual space for circulating knowledge through debating, asking, affirming, advising, and sharing experiences. As Berryman and Kavka (2018) note in their research on the creation of negative affect on YouTube, comment threads invite viewers to recirculate knowledge and experiences, using videos as affective reference points from which to generate further or alternative sites of knowing and solidarity (pp. 90–91). Although liking and commenting on videos is a standard YouTube convention, the interactions that videos facilitate between vloggers and viewers on comment threads afford further insight into the ways in which viewers respond or relate to the topic of fertility preservation.

While the number of views and types of comments in the videos analysed in our study varied depending on the vloggers' popularity, there were evident patterns in viewers' responses that suggest the perceived value of hearing others' experiences of fertility preservation. For example, viewers thanked vloggers for talking about a topic that was variously described as personal, heavy, or invasive, or for answering a 'question' which they felt they could not ask anyone else, further highlighting that fertility preservation among trans people is perceived as a taboo or sensitive topic. Many viewers sympathised with vloggers, feeling sorry for those who shared negative experiences, or commenting that they would make a great 'mum' or 'dad' one day. Just as vloggers' advice throughout their videos serves to affirm trans people's choices and self-determination, some viewers reciprocated this affirmation through their comments. Vloggers' experiential expertise highlights the ways in which viewers come to understand and 'know' fertility preservation through watching and listening to the lived experience of others.

Tensions Between Transitioning and Reproductive Futurity

Vloggers' accounts further highlight the complexities and uncertainties of decision-making around fertility preservation as a trans person. Vloggers who discussed the prospect of storing gametes commonly described their decision to do so as a complex life decision, or, as one vlogger put it, an 'uncomfortable adventure', that involved many considerations around family, identity, emotions, and the body. While some vloggers stated they were certain they wanted biogenetically related children, others shared that they felt unsure about their desire to start families in the future and the impact of delaying or stopping gender-affirming hormones, intimating that these feelings may be subject to change.

A few vloggers discussed the decision to take hormones, a medically necessary process which many trans people start in adolescence, as an important factor in making long-term decisions about their fertility. Some trans people may start taking gender-affirming hormones at a time when they are not yet certain about their reproductive futures, or do not have the financial resources or support to store their gametes. As JE (2018) mentioned, 'this is the reality of being trans, you have to make a lot of adult decisions early on for the sake of your own life and your own wellbeing'. Notably, vloggers discussed the tensions between the perceived selflessness of having children as a conventional social norm, and guilt about prioritising one's transition over storing their gametes. The uncertainty around hormones and long-term decision-making was commonly discussed among vloggers, as indicated below:

> I think even though it's going to set back the chance for me to start testosterone maybe, cos I'll have to be able to harvest and freeze my eggs. It may be a setback, but it is something that, you know, I want to do… it's a life decision. (LL, 2018)

> My transition is so much more important to me, I need to stick on my track because this is who I am, you know, I can't bring a baby into this life if I'm not me—how am I going to love something if I don't love myself? (GG, 2018)

Vloggers' negotiations regarding delaying or stopping gender-affirming hormones in order to store gametes suggest that taking hormones is inextricably tied to one's sense of self for many trans people, and is a temporal point in transition from which people move 'forward' or 'backward'. Many trans people have difficulty accessing hormones due to long wait times and high costs (Erasmus et al., 2015; Ker et al., 2020; Veale et al.,

2015; Veale et al., 2019), so postponing or pausing transition can be a significant decision to make when considering less certain reproductive futures (Chiniara et al., 2019). Some vloggers, and a few viewers who commented on the vlogs, justified their decision not to store gametes by stating that many cisgender people must work through similar uncertainties and consider other ways of starting a family, such as adoption or surrogacy. This decision-making is evident in LL's video, when they remarked:

> I've changed my mind many times about it… I don't care, some people can't have kids, it's normal, you just get donors and stuff, or whatevers [sic]… but then thinking about it, I really want my biological kids, I want them to be my kids. (LL, 2018)

Some vloggers spoke also about the logistics of storing gametes as creating further uncertainties. For example, while vloggers who were assigned female at birth spoke about the possibility of coming off testosterone and still having the capacity to reproduce, it is well documented that the process of harvesting ovarian eggs is more invasive, time-consuming, and expensive than storing sperm (see Almeling, 2011: Chap. 3). While sperm banking is more financially accessible in some healthcare systems due to government funding (such as in New Zealand), vloggers who were assigned male at birth talked about the high likelihood of becoming infertile once taking oestrogen, and therefore urged other transfeminine people to store sperm before starting hormones despite the awkwardness and discomfort of the process. Vloggers' discussions of the tensions between transitioning and their fertility, and some vloggers' uncertainty about their desire to have biogenetically related children, highlight the complex factors that trans people consider when choosing to preserve their fertility.

Mediating Intimacies Through Vlogging

As scholars have previously discussed (Chambers, 2013; Horak, 2014; Raun, 2015a, 2015b), the act of sharing and archiving bodily and affective changes on the internet can create unique ways of conceptualising and knowing the body and its possibilities. Throughout their videos, vloggers often located their bodies as sites of subjective, embodied, and affective knowledge by including 'real-time' footage of their actual or anticipated emotions, thoughts, and bodily changes in relation to the fertility preservation processes they were going through. Chambers (2013), exploring the relationship between social media and representations of intimacy, discusses the emerging social media landscape as a fluid site through which individuals are able to fashion their emotions, bodies, and thoughts on public platforms to engage with virtual communities (p. 163). To this end, vlogging affords trans people agency to share intimate details about their bodies, thoughts, and emotions while remaining physically distant when sharing and consuming information.

Vloggers who had undergone gamete storage documented the physical and emotional toll of the processes in intimate ways. A few transmasculine vloggers used metaphors to describe how their bodies looked and felt, such as their ovaries swelling to size of 'oranges' or 'tennis balls'. Many vloggers' narratives were marked by displaying their emotions or speaking about anticipated feelings. In one vlog, for example, NM (2018) showed their frustration of not injecting the right dose of synthetic hormones, began crying, and was comforted by their partner. The night before their surgery, NM spoke to the camera and expressed how much they wanted to cry. Similarly, a few transfeminine vloggers became emotional when discussing their desire to have children. Even when vloggers were not documenting the gamete storage process in real time, they often referred to how they felt, or were expecting to feel, through the process:

> I'm not gonna lie, the first I think 2 or 3 weeks were f***ing hard, I was pretty much fighting with everyone and you know, I wasn't really sure of why I was mad at everything and why I was mad at the world [...] I tried to take myself back from the situation and remind myself, you know, it's not that you hate the world, it's just your body's hating you for what you're putting it through right now. (CC, 2018)

> It is hard [...] because it's not you that's snapping, it's the hormones that make even the smallest of problems seem like somebody's just died, it's unreal. (OA, 2017)

As OA's quote demonstrates, some transmasculine vloggers attributed the emotional effects of the gamete storage process to external processes such as their hormonal or bodily changes, or being misgendered by healthcare professionals. Through unfiltered descriptions, most vloggers worked to frame emotions as an inevitable process of fertility preservation. Some vloggers virtually 'brought' viewers along with them throughout the gamete storage process, by documenting their trips to the clinic, waking up from surgery, showing viewers their medications, or doing hormone shots in real time. This footage typically showed vloggers in various emotional states, which worked to make the process more visceral to viewers.

The ways in which vloggers mediated their emotions and bodies, and therefore generated subjective knowledge about these aspects of fertility preservation, largely depended on the economic and personal contexts in which they were vlogging. In general, the videos of vloggers who had fewer subscribers to their channel presented themselves through unedited or unfiltered videos, reflecting their motivation to document their experiences of fertility preservation for themselves and/or other trans people (rather than for a larger audience). Alternatively, three 'micro-celebrity' vloggers (Raun, 2018) in our study, who produced monetised content, curated their fertility preservation experiences—while no less authentic or intimate—to appeal to a broader audience, through editing their videos in such a way that affective displays became definitive points in their narrative. As media scholars have argued, YouTubers might perform affective labour to gain subscribers, or generate revenue through advertisements (e.g. Berryman & Kavka, 2018; Raun, 2018), which may in turn affect how vloggers mediate their experiences on screen. While both groups inevitably mediated their experiences in their videos, there were notable differences to their affective displays depending on their relationship to YouTube's consumer economy.

Gender, Identity, and the Hegemony of Cisnormative Knowledge

Trans people's fertility preservation vlogs demonstrate the influence of gendered narratives in constructions of personal and social identities such as being a mother/father/parent. Many vloggers discussed negotiating the cisnormative belief that a person's status as a parent is determined by their sex assigned at birth. On the one hand, some vloggers drew on such biologically determined understandings of gender, which conflate gender and sex, when describing barriers or challenges to their own reproductive capacities. For example, TB (2018), a trans man, shared that 'I do want a family, I know that [...] but I can't do it the female way, I can't get pregnant'. Similarly, two transfeminine vloggers discussed how they hoped that the advancements in assisted reproductive technologies would enable them to bear their own children in the future, and another transfeminine vlogger described their inherent desire to have a maternal connection with their child. Many vloggers resisted being seen, legally or socially, as their child's biological mother through providing eggs, or as their father through providing sperm, due to gender dysphoria and the prospect of being misgendered. Through drawing on these gender discourses, vloggers' experiences highlight at once how the cisnormativity embedded in dominant discussions around gametes and fertility reproduce assumptions which affect trans people's sense of reproductive selfhood.

The (cis)gendering of gametes further reinforces the notion that queer and trans people's reproductive futures are less achievable or successful than cis/heteronormative reproduction. Consequently, queer and trans people may internalise dominant narratives about queer or trans possibilities, such as the idea that being queer or trans precludes one's ability to have biogenetically related children, or that queer kinship is less credible than heteronormative family formations (Von Doussa et al., 2015). For example, JE (2018) shared that, growing up queer, they 'got used to the idea' that biological reproduction was not possible for them because they could not imagine themselves in a 'stereotypical heterosexual situation'. This idea influenced JE's feeling that it was not necessary for their children to be biologically related to them.

Many transfeminine vloggers also talked about rejecting the notion of fatherhood and claiming motherhood, and some hoped that future assisted reproductive technologies (see Balayla et al., 2021) might enable them to get pregnant. This is consistent with narrative accounts of trans women's desires to carry a child via uterine transplantation (see Tonkin, 2019, pp. 91–97). Trans women's association of pregnancy with motherhood can be a way of claiming and affirming a body which society has gendered differently. Other vloggers discussed fertility preservation processes as a way of enabling them to self-determine their role as a parent, regardless of their reproductive capacities. As TTM (2019) stated, 'I can have biological children if I wanted to, even after I have my penis turned into a vagina […] there is such a thing as sperm banking'.

Vloggers further challenged normative ideals about family-building by affirming the many ways of creating families that are not predicated on biological relatedness. Almost all vloggers who had not stored their gametes, or who did not wish to do so, talked about adoption or surrogacy as equally valid alternatives to having biologically related children. Diverse ways of building families were further emphasised in viewers' comments, many of whom suggested adoption as an alternative or shared their stories about the benefits of adoption. Here, both vloggers' and viewers' comments challenge the pro-natalist assumption that being a parent is 'synonymous' with being a biological mother or father (Riggs, 2019).

Finally, people's considerations when storing gametes differ depending on their circumstances, such as time, career, and the possibility of illness or death (see Brown & Patrick, 2018; Inhorn et al., 2017; van de Wiel, 2020). An aspect unique to trans people's experiences of fertility preservation is gender dysphoria, the discomfort caused by the disconnect between one's gender and sex assigned at birth (Birenbaum-Carmeli et al., 2020; Chen et al., 2019; Erbenius & Gunnarsson Payne, 2018). Vloggers talked about how their dysphoria was triggered (or would be triggered) throughout the gamete storage process, including going off hormones, the process of retrieving gametes, and the way trans people's gender was perceived by fertility clinic staff. Of the vloggers who underwent fertility preservation, many had mixed experiences with healthcare professionals in fertility clinics. Some vloggers spoke about individual nurses and receptionist staff using their birth name, the wrong pronouns, or referring to them by their gender assigned at birth. Most vloggers discussed their experiences of misgendering as triggering their dysphoria, exacerbating what was already considered a 'dysphoric activity' (CC 2018).

In contrast, a few vloggers described the ways in which healthcare professionals who affirmed their gender had reduced the awkwardness of the gamete storage process. As existing literature indicates, trans people's positive experiences of assisted reproductive services are typically determined by healthcare professionals' use of affirming language and respect (Bartholomaeus & Riggs, 2020; James-Abra et al., 2015; Kyweluk et al., 2018), which suggests the importance of equipping healthcare professionals with training in gender diversity to mitigate trans people's dysphoria when accessing fertility preservation services.

Discussion

Online platforms such as YouTube enable trans people to produce and share intimate knowledge about fertility preservation, a topic which has until recently been largely absent from conversations about transitioning, or not talked about outside of medical settings. Vloggers' motivations to publicly document their views or experiences—namely, the lack of information elsewhere—in addition to viewers' comments on the value of such videos, indicates the crucial role that sharing experiential knowledge plays in connecting people to one another and educating trans people about fertility preservation outside of medical settings. Considering that YouTube is an increasingly popular source of information among trans communities, the knowledge trans people create through this platform undoubtedly informs viewers' understandings of, or decision-making around, their transitions and reproductive futures. Vloggers offer insight into the practical, temporal, and emotional challenges of fertility preservation, which may inform viewers' understanding of what such processes entail as a trans person.

The informality, yet perceived value, of these vlogs challenges traditional epistemic hierarchies that have historically determined what counts as 'proper knowledge' in trans healthcare discourses (Pearce, 2018, 27). Viewers' comments on the value of hearing vloggers' experiences suggest that lay knowledge and exchanges between trans people are considered equally if not more important than 'facts' or scientific knowledge to many trans people. Further, vloggers discussed trans-specific issues such as dysphoria and misgendering, in ways that assumed viewers' understanding and relation to these concepts and thus created in-group knowledge. As Fazey et al. (2006) note, both experiential knowledge and more objective information, such as location-specific or scientific knowledge, can inform how people understand certain phenomena. As patients' lived experience and community collaboration begin to influence the development of informed consent models in trans healthcare, as is the case in New Zealand (Johnson & Oliphant, 2019), content produced by and for lay trans people will have an increasingly important role to play in shaping not only trans people's understanding of what is possible for them, but also healthcare professionals' understandings of trans people's experiences.

The information shared through YouTube videos can transform the ways in which trans people understand themselves (Raun, 2010). Similarly, vloggers' lived experiences of fertility preservation may have the potential to inform healthcare professionals' engagement with trans people considering or undergoing fertility preservation. The issues documented in our study are consistent with barriers to fertility preservation outlined in existing literature, such as the use of gendered language and assumptions among healthcare professionals (Epstein, 2018; James-Abra et al., 2015; Riggs, 2019; Riggs & Bartholomaeus, 2018), and trans people's varying desire for children biogenetically related to them (Chen et al., 2019; Von Doussa et al., 2015). The diversity across trans vloggers' experiences and levels of certainty around storing gametes underscores the importance of questioning assumptions that everyone will want biological children, or that having biogenetically related children is the only way to start a family (Riggs, 2019). At the same time, these findings support the need for accurate information about fertility preservation at any stage of their transition, to increase trans people's self-determination when making decisions about their reproductive futures.

As fertility preservation is typically discussed through a cisnormative lens, it is important to consider the material impacts that such discourses have on how trans people relate to the notion of reproduction and gamete storage. For example, most vloggers considered how decisions around their reproductive futures were influenced by how masculine- or feminine-specific spaces and processes felt to them. While some vloggers talked about the perception of being seen as the child's 'father' or 'mother' based on their sex assigned at birth as a barrier to storing their gametes, other vloggers rejected the notion that parenthood was defined by their biology, thus queering the notion of reproduction and parenthood. Some vloggers' accounts of being misgendered by healthcare professionals at fertility clinics, or presenting their gender in a certain way to avoid being misgendered, highlight the cisnormative assumptions that many healthcare professionals carry. Viewers' comments on the lack of trans-specific information in fertility discourses suggest that vloggers' affective and embodied representations create possibilities for trans people to imagine their reproductive futures beyond cisnormative discourses of fertility and the body. These experiences point to the need for emerging fertility preservation discourses to frame families, identities, and reproduction in ways that account for the diversity of and complexities between gender and sex.

Conclusion

This exploratory study has sought to understand the different types of knowledge that trans vloggers produce, negotiate, and share through publicly available YouTube vlogs. The vloggers in our study created and shared experiential knowledge by sharing their own thoughts and experiences, positioning themselves as lay people but experts of their own experiences. They set the limits of their experiential knowledge by discussing the challenges and tensions between transitioning and planning for their reproductive futures. Vloggers' intimate presentations of their emotions and bodies may bring viewers 'closer' to their experiences and present viewers with possibilities otherwise not included in online information or clinical settings. Vloggers' videos offer further insight into trans people's experiences of cisnormative knowledge and gender dysphoria in relation to gamete storage processes. These experiences reveal the cisnormativity in fertility services and discourses, the challenging of which will have positive impacts on trans people's experiences of accessing fertility preservation. Importantly, our findings suggest there is both epistemic value and potential in vloggers' experiential expertise for trans people considering fertility preservation, and for healthcare providers wanting to better understand the unique needs of this diverse and historically underserved population.

Acknowledgements Thank you to the vloggers for sharing their experiences online and generously giving their consent for us to use quotes to illustrate our findings. Rhonda Shaw would also like to thank Victoria University of Wellington's FHSS Joint Research Committee for funding the *Reproductive Futures* study (Grant: 224801), and Alex Ker for his research assistance on this project.

Appendix: Inclusion Criteria

YouTube videos were included if:

- The video was made by a trans person, about their personal experiences or views of fertility preservation
- Vloggers were 18 years or older at the time of creating the video
- The video was in English.

There were no criteria around the date the video was uploaded, the length of the video, or number of viewers or followers a vlogger had.

Notes

1. In this chapter, we use the umbrella term 'gamete storage' to include both ovarian egg and sperm freezing, the processes commonly undertaken by trans vloggers in our study. Ovarian egg preservation, or egg freezing, entails injecting synthetic hormones, sometimes known as 'trigger shots', to stimulate the ovarian follicles. These hormones, typically taken over a two-week period, can affect a person's mood and cause bloating or physical discomfort. The person then undergoes a procedure to obtain mature eggs (under sedation), which are frozen for future use. Sperm freezing most commonly involves masturbation to obtain sperm, and is considered less medically invasive and resource-intensive than ovarian egg freezing. Gamete storage involves the cryopreservation of ovarian eggs or sperm at sub-zero temperatures of minus 196 °C. Cryogenic storage at these temperatures provides a near indefinite longevity to the preserved cells, assuming samples are safe and checked regularly. Legislation in most jurisdictions,

however, regulates the length of time gametes may be stored for. In Aotearoa New Zealand, for example, there is a ten-year storage limit before a request for an extended period must be made.
2. While we use 'trans' in this chapter to include transgender and non-binary genders, we acknowledge that not all people use nor relate to this term.
3. Medical transition refers to accessing gender-affirming healthcare, or medically necessary treatments and processes many (but not all) trans people undertake to affirm their gender. These processes include (but are not limited to) taking gender-affirming hormones, puberty blockers, laser hair removal, chest reconstruction surgery, and genital reconstruction surgery.
4. See Chap. 5 in this volume for further discussion of this point.
5. Aotearoa New Zealand has been referenced here because this is where the authors are based. We acknowledge that fertility clinics in other countries are increasingly recognising the importance of ensuring diverse representations of gender and sexuality on their media platforms.

References

Akrich, M. (2010). From communities of practice to epistemic communities: Health mobilizations on the internet. *Sociological Research Online, 15*(2), 116–132.

Akrich, M., Nunes, J., & Paterson, F. (2008). *The dynamics of patient organizations in Europe* (pp. 13–82). Presses des Mines.

Almeling, R. (2011). *Sex Cells: The medical market for eggs and sperm*. University of California Press.

Balayla, J., Pounds, P., Lasry, A., Volodarsky-Perel, A., & Gil, Y. (2021). The Montreal Criteria and uterine transplants in transgender women. *Bioethics, 35*(4), 326–330.

Bartholomaeus, C., & Riggs, D. (2020). Transgender and non-binary Australians' experiences with healthcare professionals in relation to fertility preservation. *Culture, Health & Sexuality, 22*(2), 129–145.

Berryman, R., & Kavka, M. (2018). Crying on YouTube: Vlogs, self-exposure and the productivity of negative affect. *Convergence, 24*(1), 85–98.

Birenbaum-Carmeli, D., Inhorn, M. C., & Patrizio, P. (2020). Transgender men's fertility preservation: Experiences, social support, and the quest for genetic parenthood. *Culture, Health & Sexuality, 23*(7), 945–960.

Brown, E., & Patrick, M. (2018). Time, anticipation, and the life course: Egg freezing as temporarily disentangling romance and reproduction. *American Sociological Review, 83*(5), 959–982.

Chambers, D. (2013). *Social media and personal relationships: Online intimacies and networked friendship*. Springer.

Chen, D., Kyweluk, M., Sajwani, A., Gordon, E., Johnson, E., Finlayson, C., & Woodruff, T. (2019). Factors affecting fertility decision-making among transgender adolescents and young adults. *LGBT Health, 6*(3), 17–115.

Chen, D., Matson, M., Macapagal, K., Johnson, E. K., Rosoklija, I., Finlayson, C., Fisher, C. B., & Mustanski, B. (2018). Attitudes toward fertility and reproductive health among transgender and gender-nonconforming adolescents. *The Journal of Adolescent Health: Official Publication of the Society for Adolescent Medicine, 63*(1), 62–68.

Chiniara, L. N., Viner, C., Palmert, M., & Bonifacio, H. (2019). Perspectives on fertility preservation and parenthood among transgender youth and their parents. *Archives of Disease in Childhood, 104*, 739–744.

Dewey, J. M. (2008). Knowledge legitimacy: How trans-patient behavior supports and challenges current medical knowledge. *Qualitative Health Research, 18*(10), 1345–1355.

Eckstein, A. J. (2018). Out of sync: Complex temporality in transgender men's YouTube transition channels. *QED: A Journal in GLBTQ Worldmaking, 5*(1), 24–47.

Ellis, S. J., Bailey, L., & McNeil, J. (2015). Trans people's experiences of mental health and gender identity services: A UK study. *Journal of Gay & Lesbian Mental Health, 19*(1), 4–20.

Epstein, R. (2018). Space invaders: Queer and trans bodies in fertility clinics. *Sexualities, 21*(7), 1039–1058.

Erasmus, J., Bagga, H., & Harte, F. (2015). Assessing patient satisfaction with a multidisciplinary gender dysphoria clinic in Melbourne. *Australasian Psychiatry, 23*(2), 158–162.

Erbenius, T., & Gunnarsson Payne, J. (2018). Unlearning cisnormativity in the clinic: Enacting transgender reproductive rights in everyday patient encounters. *Journal of International Women's Studies, 20*(1), 27–39.

Fazey, I., Fazey, J., Salisbury, J., Lindenmayer, D., & Dovers, S. (2006). The nature and role of experiential knowledge for environmental conservation. *Environmental Conservation, 33*, 1–10.

Giles, D. (2017). Online discussion forums: A rich and vibrant source of data. In V. Braun, V. Clarke, & D. Gray (Eds.), *Collecting qualitative data: A practi-*

cal guide to textual, media and virtual techniques (pp. 189–210). Cambridge University Press.

Horak, L. (2014). Trans on YouTube: Intimacy, visibility, temporality. *Transgender Studies Quarterly, 1*(4), 572–585.

Hsieh, H. F., & Shannon, S. E. (2005). Three approaches to qualitative content analysis. *Qualitative Health Research, 15*(9), 1277–1288.

Inhorn, M. C., Birenbaum-Carmeli, D., & Patrizio, P. (2017). Medical egg freezing and cancer patients' hopes: Fertility preservation at the intersection of life and death. *Social Science & Medicine, 195*, 25–33.

James-Abra, S., Tarasoff, L. A., Green, D., Epstein, R., Anderson, S., Marvel, S., Steele, L. S., & Ross, L. E. (2015). Trans people's experiences with assisted reproduction services: A qualitative study. *Human Reproduction, 30*, 1365–1374.

Johnson, R., & Oliphant, J. (2019, May). *Gender-affirming care pathways in Aotearoa. Presentations and panel discussion facilitated by Dr Rachel Johnson and Dr Jeannie Oliphant, including speakers from the following DHBs: Auckland/Waitemata/Counties Manukau, Waikato, Lakes, Hawke's Bay, Mid Central, Capital and Coast/Hutt Valley/Wairarapa, Nelson Marlborough, and Canterbury.* Paper Presented at Aotearoa New Zealand Trans Health Symposium, University of Waikato, New Zealand. https://www.ivvy.com.au/event/LCQCA2/presentations.html

Ker, A., Fraser, G., Lyons, A., Stephenson, C., & Fleming, T. (2020). Providing gender-affirming hormone therapy through primary care: Service users' and health professionals' experiences of a pilot clinic. *Journal of Primary Health Care, 12*(1), 72–78.

Kyweluk, M., Sajwani, A., & Chen, D. (2018). Freezing for the future: Transgender youth respond to medical fertility preservation. *International Journal of Transgenderism, 19*(4), 401–416.

Madill, A., Jordan, A., & Shirley, C. (2000). Objectivity and reliability in qualitative analysis: Realist, contextualist and radical constructionist epistemologies. *British Journal of Psychology, 91*(1), 1–20.

Mamo, L. (2013). Queering the fertility clinic. *The Journal of Medical Humanities, 34*, 227–239.

Markham, A. (2012). Fabrication as ethical practice: Qualitative inquiry in ambiguous Internet contexts. *Information, Communication & Society, 15*(3), 334–353.

Miller, B. (2017). YouTube as educator: A content analysis of issues, themes, and the educational value of transgender-created online videos. *Social Media & Society, 3*(2), 1–12.

Murphy, T. F. (2012). The ethics of fertility preservation in transgender body modifications. *Bioethical Inquiry, 9*, 311–316.

Nebeling Petersen, M., Harrison, K., Raun, T., & Andreassen, R. (2018). Introduction: Mediated intimacies. In R. Andreassen, M. Nebeling Petersen, K. Harrison, & T. Raun (Eds.), *Mediated intimacies: Connectivities, relationalities and proximities* (pp. 1–16). Routledge.

O'Neill, M. G. (2014). Transgender youth and YouTube videos: Self-representation and five identifiable trans youth narratives. In C. Pullen (Ed.), *Queer youth and media cultures* (pp. 34–45). Palgrave Macmillan.

Patterson, A. N. (2018). YouTube generated video clips as qualitative research data: One researcher's reflections on the process. *Qualitative Inquiry, 24*(10), 759–767.

Pearce, R. (2018). *Understanding trans health: Discourse, power and possibility*. Policy Press.

Poteat, T., German, D., & Kerrigan, D. (2013). Managing uncertainty: A grounded theory of stigma in transgender health care encounters. *Social Science & Medicine, 84*, 22–29.

Radi, B. (2019). On trans* epistemology: Critiques, contributions, and challenges. *Transgender Studies Quarterly, 6*(1), 43–63.

Raun, T. (2010). Screen-births: Exploring the transformative potential in trans video blogs on YouTube. *Graduate Journal of Social Science, 7*(2), 113–130.

Raun, T. (2015a). Video blogging as a vehicle of transformation: Exploring the intersection between trans identity and information technology. *International Journal of Cultural Studies, 18*(3), 365–378.

Raun, T. (2015b). Archiving the wonders of testosterone via YouTube. *Transgender Studies Quarterly, 2*(4), 701–709.

Raun, T. (2018). Capitalizing intimacy: New subcultural forms of microcelebrity strategies and affective labour on YouTube. *Convergence, 24*(1), 99–113.

Riggs, D. (2019). An examination of the "just in case" arguments as they are applied to fertility preservation for transgender people. In V. Mackie, N. J. Marks, & S. Ferber (Eds.), *The reproductive industry: Intimate experiences and global processes* (pp. 69–78). Rowman and Littlefield.

Riggs, D. W., & Bartholomaeus, C. (2018). Fertility preservation decision making amongst Australian transgender and non-binary adults. *Reproductive Health, 15*(1), 181–190.

Stroumsa, D., Shires, D. A., Richardson, C. R., Jaffee, K. D., & Woodford, M. R. (2019). Transphobia rather than education predicts provider knowledge of transgender health care. *Medical Education, 53*(4), 398–407.

Tattersall, R. (2002). The expert patient: A new approach to chronic disease management for the twenty-first century. *Clinical Medicine, 2*(3), 227.

Tonkin, L. (2019). *Motherhood missed: Stories from women who are childless by circumstance.* Jessica Kingsley Publishers.

Turner, B. S. (2008). Citizenship, reproduction and the state: International marriage and human rights. *Citizenship Studies, 12*(1), 45–54.

van de Wiel, L. (2020). *Freezing fertility: Oocyte cyropreservation and the gender politics of aging.* New York University Press.

Vance, S. R., Halpern-Felscher, B. L., & Rosenthal, S. M. (2015). Health care providers' comfort with and barriers to care of transgender youth. *Journal of Adolescent Health, 56*(2), 251–253.

Veale, J., Byrne, J., Tan, K., Guy, S., Yee, A., Nopera, T., & Bentham, R. (2019). *Counting ourselves: The health and wellbeing of trans and non-binary people in Aotearoa New Zealand.* Transgender Health Research Lab, University of Waikato: Hamilton NZ. https://countingourselves.nz/wp-content/uploads/2019/09/Counting-Ourselves_FINAL.pdf

Veale, J., Saewyc, E., Frohard-Dourlent, H., Dobson, S., Clark, B., & The Canadian Trans Youth Health Survey Research Group. (2015). *Being safe, being me: Results of the Canadian trans youth health survey.* Stigma and Resilience Among Vulnerable Youth Centre, School of Nursing, University of British Columbia.

Von Doussa, H., Power, J., & Riggs, D. (2015). Imagining parenthood: The possibilities and experiences of parenthood among transgender people. *Culture, Health & Sexuality, 17*(9), 1119–1131.

Wehling, P., Viehöver, W., & Koenen, S. (2015). Patient associations, health social movements and the public shaping of biomedical research: An introduction. In P. Wehling, W. Viehöver, & S. Koenen (Eds.), *The public shaping of medical research: Patient associations, health movements and biomedicine* (pp. 1–20). Routledge.

Wu, H., Yin, O., Monseur, B., Selter, J., Collins, L., Lau, B., & Christianson, M. (2017). Lesbian, gay, bisexual, transgender content on reproductive endocrinology and infertility clinic websites. *Fertility and Sterility, 108*(1), 183–191.

5

Fertility and Fragility: Social Egg Freezing and the 'Potentially Maternal' Subject

Julie Stephens

Introduction

At the very time when the freezing of human reproductive tissue promises a world of reproductive immortality, the finitude of the human and natural world has never been more palpable. The death and destruction caused by rising temperatures, bushfires, floods, and the global pandemic stand in stark contrast to notions of indefinite fertility without bodily limitations or temporal constraints. The rhetoric around assisted reproduction that once emphasised biological limits in the face of age-related infertility has been given a new inflexion in biotechnologies concerned with fertility preservation and extension. The feared image of the 'biological clock' has been replaced by a kind of 'Tardis', a time machine storing frozen reproductive material in a state of suspended animation ready to be brought to life when social circumstances change. In the current moment, notions of biological limits have been replaced by

J. Stephens (✉)
Victoria University, Melbourne, VIC, Australia
e-mail: julie.stephens@vu.edu.au

indeterminate promises of endless fertility preservation or what Lucy van de Wiel calls a 'bioprepared fertility' (Van de Wiel, 2015, p. 126). The suspension of fertility in time has produced new subject positions: 'potentially infertile' and 'potentially maternal'. It is at this point that biotechnological imaginaries intersect with other discourses.

This chapter is concerned to critique some of these intersecting points, where fertility preservation as an assisted reproduction technology is depicted as the solution to a social problem and a logical extension of the project of mainstream feminism. A range of cases of 'social freezing' could be relevant to this investigation. The freezing of ovarian tissue and its experimental transplantation into menopausal women would be a recent example because it has been promoted as a possibility of delaying menopause so women can concentrate on building their careers by having children very much later in life (Sample, 2019). The figure of the 'postmenopausal mother', which Nolwenn Bühler characterises as the 'making of older mothers' (Bühler, 2015, p. 70), could be seen as being consistent with second-wave feminism's refusal for biology to be destiny. However, my attention here is less on the prospect of an 'ageless motherhood' (Bühler, 2021) and more on the content or lack of it in how the potentially maternal is conceptualised. The focus will be on representations of social egg freezing; the harvesting and preservation of a woman's ovarian eggs in order for the possibility of childbearing to be deferred to a future time when it is more conducive to personal and/or professional circumstances. Aspirational maternity, in this regard, carries with it a set of liberal feminist assumptions about empowered mothers and motherhood as one path among many for women to realise an unlimited, individually-centred potential.

At the outset, a word on terminology. There are contests around all terms in the phrase 'social egg freezing for non-medical purposes'. In an industry where age itself is medicalised, egg-freezing or oöcyte cryopreservation is viewed as a 'preventative medical procedure' rather than a socially embedded practice. Dominic Stoop and fellow authors argue that egg freezing is neither 'social' nor 'non-medical' and suggest the name of the procedure should be revised to 'AGE banking' (oöcyte banking for anticipated gamete exhaustion) (Stoop et al., 2014). The fertility industry prefers to call the technique 'elective' egg freezing, emphasising 'choice' and downplaying the social factors that may have contributed to delaying childbearing or the socio-economic barriers preventing most women from affording to 'elect' to freeze their eggs.

Lucy van de Wiel warns against reproducing an opposition between medical and social motivations for egg freezing, with one being controversial and the other, not. She maps the way these oppositions can anchor moralising narratives about reproductive decision-making (Van de Wiel, 2014, p. 4). Yet to downplay social factors would be to risk reproducing another binary between 'external' social formations and so-called internal psychic life. It would ignore the fact that our everyday, intimate, and psychic lives are indivisible from our histories, group identifications, dominant cultural narratives, and our networked identities. Moreover, it would disguise the clear socio-economic dimensions of the fertility preservation industry. Alessandra Alteri and colleagues record that the procedure to date is commonly used by 'Caucasian, highly educated, middle-class professional women in their mid-to late 30s'(Alteri et al., 2019, p. 647). Other studies confirm that oöcyte banking is only accessible to a small group of highly privileged women in the global North (Petropanagos et al., 2015). For these reasons, and others that will be elaborated below, it is important to retain the term 'social' in reference to the complex desires for and uses of the technology of egg freezing.

Before investigating the overlapping and often contradictory discourses around fertility preservation, two further areas of unease need to be mentioned briefly. The first is the absence, in some critical analyses of an attempt to theorise or include the despair and loss that is felt by those unable to have a child. These approaches risk appearing abstract and partial. There is a danger that reproductive desire or the consumption of fertility services become represented as a kind of false consciousness and that the industry would disappear if only people knew its exploitative side.[1] Evidence shows that clearly this is not the case, as attested by the continued expansion of the fertility industry, where 'babymaking' purportedly 'has become big business' with capital investment skyrocketing in the sector (Kowitt, 2020). While critiques of the industry go back to the 1980s, the motivations to reproduce are far more complicated and elusive than ever can be captured adequately by conventional social science approaches. This is best expressed in the words of Nancy Chodorow, who shows that the desire for a child, like the desire to delay maternity, is based on a constellation of fantasies and defences and often intra-psychic conflict. She argues that we attend to the complexity of this desire which begins 'internally in the conflictual, intense cauldron of childhood sexuality and object relations, and is overdetermined [and] filled with fantasy' (Chodorow, 2003, p. 1181). Psychoanalytic and some psychosocial approaches touch on the dense layers of affect and meaning that persist, even when that desire is never fulfilled.[2]

By contrast, the breadth of current developments in the global biotechnology industry is well captured variously as the 'tissue economy', 'biocapital', the 'bioeconomy', or 'capital mobilising the biological', to name but a few of the most evocative designations (Weinbaum, 2019). Rhonda Shaw maps the theoretical shift towards various 'bio-concepts and bioeconomics' as ways to examine the transfer and circulation of human biological material, such as cells, tissues, and bodily fluids as transformed objects of capital exchange (Shaw, 2017). According to Melinda Cooper, the biotech revolution is 'the result of a whole series of legislative and regulatory measures designed to relocate economic production at the genetic, microbial, and cellular level, so that life becomes, literally, annexed within capitalist processes of accumulation' (Cooper, 2008, p. 19). The global dimensions of the processes in which the most 'intimate, embodied elements of biological reproduction are embedded' have also been well explored in different transnational studies (Mackie et al., 2019), especially those with a focus on the Asia-Pacific region. Alongside the excellent work on the commodification and racialisation of intimate *in vivo* labour in expanding areas such as commercial surrogacy, there have been many compelling arguments about reproductive justice, who has and does not have access to these developments, reproductive autonomy (Harwood, 2009), gift narratives, and the ethical dilemmas encountered in such a contested area. However, even ethnographic studies or those revolving around qualitative interviews can fall short as far as tapping into the deepest embedded meanings and fantasies around genetic motherhood. This is a point I will return to later.

The second area of concern is that there is little published research focusing on transgender and non-binary people who use egg freezing to preserve fertility prior to transitioning and surgical or hormone therapy.[3]

Damien Riggs and Clare Bartholomaeus' pioneering work in this area argues that if genetic relatedness is viewed as important by transgender men and women, then the availability of this technology becomes a reproductive rights issue (Riggs & Bartholomaeus, 2018). Elsewhere, Riggs, who is an Australian-based researcher working across social science disciplines from critical kinship studies to psychotherapy, draws attention to a slippage between the desire to preserve one's fertility (or the 'just in case' advice given to those transitioning) and a desire to have a child. He argues for a clinical practice that reflects the complexities of decision-making about reproduction and underlines that it is important not to reinforce a pronatalist logic and normative assumptions about parenting (Riggs, 2019). This slippage seems to be present in the egg freezing literature in general, and little research seems to have been done on how the desire to preserve one's fertility may be very different to wanting to have a genetically related child. The former does not necessarily lead to the latter. In the case of those in the transgender and gender-diverse community, only a very small number of Riggs' participants wanted to have children to whom they were genetically related (Riggs, 2019). There seemed to be a greater recognition of the many paths to parenting and less fixation on genetic parenthood than in other groups.

Following the dominant literature on fertility preservation, my references will be to women who identify as female in their use of social egg freezing. This is not only due to the lack of research on transgender men but also that it could be argued that preserving fertility before transitioning could be construed as 'medical egg freezing' rather than 'social egg freezing', although as I have indicated above, the one cannot easily be separated from the other. What Riggs' research highlights so well is that even though the biomedical rhetoric around fertility preservation reads as socially progressive and inclusive, its application often comes with normative, ambiguous, and conservative assumptions about parenthood and family forms.

The concept of 'neoliberalism' is a term often employed by those critical of the normative assumptions about women and career in the egg-freezing debate. I will refer to neoliberalism as a dominant cultural logic, as much as a set of economic and social policies or political ideologies. This logic emerged from the way liberal doctrines of individual responsibility, small government, and a self-regulating market resurfaced in the 1980s and were then transformed to have a new global inflexion and reach (see Harvey, 2007). My attention will be on the changed understandings of the self that have accompanied neoliberal social restructuring; in short, the self as an ideal, entrepreneurial, unencumbered, and invulnerable individual. I will argue that these changed understandings have also been reproduced in mainstream liberal feminism and leave little room for ideas of motherhood as anything other than a form of self-actualisation. The maternal therefore risks becoming little more than a vague 'potentiality', ever projected to an indistinct, future time.

Banking Eggs in the Neoliberal Refrigerator

It is necessary to document a shift in the meaning and use of oöcyte cryopreservation that has followed a fairly recent change in the technology of egg freezing. The process was initially developed as a medical procedure to preserve some of the eggs of a woman prior to chemotherapy treatment or other potentially damaging medical interventions (see Kato, 2016). In such a context, one imagines that this technique was used when a woman may have had less agency than at in any other time in her life and where time was possibly short. It was not linked to a feminist idea of empowerment for women. Rather, it occurred in situations where there was little or no 'choice' in the neoliberal sense of the term where a self-sufficient, unencumbered individual makes unrestrained, rational, 'market-based' decisions. Another feature of the medical egg-freezing procedures of the past was that a woman's eggs could be damaged by the process itself, if the water in human egg cells crystallised during freezing. As such, it was an unreliable form of insurance to ensure genetic motherhood. This changed, as did the social vocabularies around egg freezing, when a new technique called vitrification was developed that freezes human eggs so rapidly that crystals do not form and hence the eggs are better preserved (Argyle et al., 2016).

It is difficult to pinpoint the specific moment in the development of this innovation in cellular freezing when medical discourses became complexly intermingled with something more contradictory and laden with a whole set of implicit directives about living a healthy lifestyle, ageing, normative motherhood, risk, choice, empowerment, technology, and most importantly, time management. Elements of the discourse concerning fertility preservation constitute what Anna Sofie Bach and Charlotte Kroløkke call a new 'biomedical imaginary' (Bach & Kroløkke, 2020). It can be detected in the scholarly literature, the promotional material generated by the reproductive industry and in popular cultural debate. Note, for instance, the way the idea of investing the future becomes a key point of reference in discussions of egg freezing as a form of 'fertility insurance'. An example from an online biomedical journal illustrates the way an attempt to introduce an affective dimension to the issue easily slips into market terminology. While the authors are cautious about whether there is an established link between a woman freezing her eggs in a timely manner and the birth of a live baby, they nonetheless resort to a non-medical comparison that 'those frozen eggs offer a chance of achieving her heart's desire. After all, we do not counsel people against taking out fire insurance because the chance of their house catching fire is very low' (Lockwood & Fraser, 2018, p. 383). This coupling of the prevention of biological ageing of a woman's eggs with protecting a house from catching on fire illustrates the extent to which commercial logics have become commonplace and have infiltrated areas once considered private and intimate.

So, even if we were to reproduce the biomedical industry's use of the term 'insurance', the jury seems to be out as to the effectiveness of oöcyte cryopreservation in improving the chances of getting pregnant and having a genetically related baby at a later stage in life. Some clinics make this clear in their advertising. Others blur the lines somewhat, and there is a disjunct between the science and the marketing. In those cases, the process is not represented as an insurance policy but more as one where women can demonstrate their agency by being proactive about protecting fertility during a 'reproductive window'. The technology is promoted as 'preserving the opportunity to have children' (Adora Fertility, 2020). Kylie Baldwin has done some of the most thorough and nuanced research on this area, detailing the way fertility is currently being constructed:

> The notion of egg freezing as a form of consumption is evident in the language used in its discussion, which is laden with market-based and eco-

nomic references and metaphors. Clinics, advertisers, as well as users of the technology, commonly describe egg freezing as an investment, and as a form of insurance, as something they hope will buy them time should they bank enough eggs. This language highlights how fertility is constructed as a commodity bought or invested in, in order to ensure a more favourable outcome for the user in the pursuit of self-actualisation. (Baldwin, 2019, p. 272)

Her interviews with users of the technology highlight the extent to which the insurance metaphor continues to resonate, even when women might be presented with contradictory or conflicting evidence about the success rates for egg freezing in preserving fertility. Nonetheless, egg freezing continues to be the preferred form of fertility preservation for women. In the Australian state of Victoria alone, data collected by the Victorian Assisted Reproductive Treatment Authority (VARTA) shows that the number of women with eggs in storage has doubled in Victoria over the last four years.[4] To return to an earlier point, the reasons are social, not medical. Interestingly, while delaying the decision to have children is often portrayed as necessary for women to build a career, the most documented reason for freezing one's eggs is either waiting to find the right partner or having a partner who is unwilling to commit to having a child (Pritchard et al., 2018).

Before discussing the important psychosocial research that Baldwin has done with her colleague Lorraine Culley on this issue (Baldwin & Culley, 2020), it is worth considering why, in the popular imagination, egg freezing is depicted as a wise career move for ambitious women. The emergence of publicity about egg freezing dates from around 2014. Much debate was prompted by a cover of *Bloomberg Businessweek* with the headline 'Freeze your eggs and free your career'. A young female executive was photographed in a defiant stance accompanied by the statement that 'now there is a new fertility procedure giving women more choices in the quest to have it all'.[5] The announcement by Apple and Facebook to provide financial support for their female employees to access fertility services as part of their corporate commitment to employee health was widely applauded. Alongside gym membership and healthy-eating plans, corporations began offering financial assistance for surrogacy, adoption,

5 Fertility and Fragility: Social Egg Freezing and the 'Potentially...

and of course, social egg freezing. Young women on a leadership track were provided with the opportunity to delay having children beyond their most fertile years and given a large stipend to assist with the cost of freezing their eggs, if that was the option they chose. This provision, framed as a commitment to diversity, remains a key plank in the business plans of the giant tech corporations.

In the public debate that followed the Apple Corporation's announcement about this new health and fertility option for female employees, commentary ranged from celebrating the opportunity egg freezing provided for women by giving them more control over decisions about reproduction, to concern about the company interfering in the personal decisions of professional women. Both sides of this debate were mounted around the idea of choice. Both also signalled important questions about market relationships and where, if ever, the market stops and intimate, private life begins. Nevertheless, mainstream liberal feminist arguments that position oöcyte cryopreservation as a rational choice for career purposes seem to dominate popular discussion. The frequent use of recognisable feminist idioms is more and more central to this configuration. Phrases include 'increasing female empowerment and control' and 'more choice for women'. Notably, a particular version of feminism is at play here. 'Choice' is the operative word referring to individual choice as a path to self-empowerment. The act of choosing—whether it be the right career or the right gym—becomes a form of self-actualisation and curiously is equated with being a feminist. No mention is made in the egg freezing discourse of other kinds of feminism that call for systemic radical change or collective action.

The dominance of the mainstream feminist view is reflected in the fact that at the same time as the giant tech corporations were announcing their 'fertility packages' for women, 'Eggs in the City' events were being held in Melbourne, Australia. These ostensibly educational occasions run by fertility centres attracted what observers called 'typical freezers': educated, successful, and driven women from their late 20s to early 40s (Fyfe & Davies, 2015). Referencing the outspoken female characters in the popular television drama *Sex and the City* (a programme about buying shoes, looking for Mr Right, and preserving one's viability on the sexual market), these events have a global reach, with one that was scheduled in

San Francisco as late as in 2019 (Eventbrite, 2020). Strategic employment of this variety of feminism is a crucial factor linking the harvesting and exchange of biological material to the processes of capitalist accumulation. In the *Fortune Magazine* article that promotes the opportunities for investment in fertility technologies, cited above, there is a discussion of 'boutique egg freezing studios' and the startups that attempt to distinguish themselves from 'infertility clinics'. 'Kindbody' is one that advertises itself as a 'women's health network'. In an event billed as *Fertility* 101 that took place in November 2019, the aesthetic is described as 'spa-like' because the CEO does not think 'health care has to be ugly'. While attendees were told to 'grab a glass of prosecco, get some cheese and relax', the wall mural behind them proclaims, 'Own Your Own Future' (Kowitt, 2020).

This individualised 'feminism lite' revolving around ideas of personal choice and empowerment also envelopes online commentary on this topic. The 'power' of egg-banking is even endorsed as a form of stress alleviation (Shady Grove Fertility, 2020). In online posts at the time of Apple's announcement about the company's fertility policies, social freezing was depicted as a wonderful option providing women with choice, freedom, and control. The purported 'empowerment' of being able to delay a family through biobanking seems to have overshadowed discussion about the structural aspects in workplaces which make parenting difficult for parents of young children in general, and mothers in particular. Similarly, internal struggles over the decision to have a child are reduced to questions of time management with fertility decline as the focus. This displaces wider emotional concerns about motherhood and what being a mother might mean for a person, alone or with a partner.

Baldwin and Culley attempt to document the emotional terrain women negotiate when undertaking the various steps in the egg-freezing process. In thematic interviews with 31 women, the first hurdle seemed to be obtaining reliable or consistent published data about a clinic's success rate regarding the number of live births achieved from frozen eggs. This meant that women had to rely on a variety of sources which included news and media stories and websites. The struggle to get 'clinic-specific success rates stratified by age' (Baldwin & Culley, 2020, p. 190) seemed to be compounded by the conflicting feelings women reported to having had at the clinics themselves. Far from the image promoted of young, independent, single, and joyously smiling women projecting a wise confidence about their decision, the participants in Baldwin and Culley's study reported not only the physical discomfort of the actual harvesting process but also the emotional pain they felt in going through it alone, without a partner. This included 'crying [their] eyes out' while waiting on a gurney before going into the operating theatre or feeling like a failure about not having a partner, or embarrassment and discomfort in front of clinic staff. In the words of one participant: 'I would just come home from every single one of those appointments and just sob. I'd just cry, just ball my eyes out, because it was such a reminder of the fact that I'm not, you know, I wasn't where I wanted to be in my life' (Baldwin & Culley, 2020, p. 189). As well as the value of recording these experiences, this important study raised another question about the idea of insurance and the ways it is currently used in the scholarly, clinical and media literature on this topic. The women in this study seemed to be insuring themselves against something far more elusive than simply increasing their chances of achieving genetic motherhood. Oöcyte freezing was a protection against 'future feelings of regret' and self-reproach about not having done enough to ensure they can have a child. This evidence is supported by other research and captured so well in the title of Bach and Kroløkke's article 'Hope and Happy Futurity in the Cryotank' (Bach & Kroløkke, 2020).

Clearly, there is more stored in the biobank than just human tissue. The desire to protect oneself from future regret is what Joanna Radin calls a form of 'planned hindsight' (Radin, 2015). It is a way to act in the present to organise and prevent future emotional pain and discomfort. While the biological material that is preserved may indeed be considered as latent or 'suspenseful matter', it is capable of producing new meanings and social relations.

It is imperative to note that the women interviewed by Baldwin and Culley did not (at the time of interview) regret going through the harvesting and freezing process despite the discomfort and indignity they experienced. They remained optimistic about a positive outcome for their eggs or believed that by the time they were ready to become pregnant, the technology in the biomedical industry would have improved. What is significant about this study is the challenge to the neoliberal and often media-charged idea that this is just one option among many for savvy, empowered, professional women. During the debate mentioned earlier concerning corporations instituting egg-freezing policies, countless representations at the time reproduced the idea that this was just a simple, logical choice. One commentator described it as 'a pretty enlightened option', just another offering or tool women can use, like the gyms some companies provide, or perks like Hawaiian vacations, salary sacrifice or even a car park (Rosewarne, 2014). The emotional challenges women may face in making the decision to bank their eggs or the intrusive nature of the medical procedures they have to endure are completely overshadowed in this discourse about choice. It should go without saying that decisions about motherhood and fertility differ, in every possible way, from deciding whether or not to take advantage of the company's physioservice or the on-site laundry facility or opting for a company car park. In a neoliberal context, there are ideological reasons that these differences need to be ironed out, primarily to create the appearance that these new tendencies in capitalist commodification are just an extension of previous or current arrangements with a comparable affect attached to them. As the anxiety and anguish of the women interviewed attest, the actual experience of the process cannot always be comfortably assimilated through this narrow ideological lens.

Technological Utopianism

Both in scholarly research and in popular online discussion, women signing up to fertility preservation through egg freezing appear well informed and realistic about the low chances their stored eggs will lead to a live birth. Qualifying the notion of insurance, they use phrases like 'not 100% certain' or 'no guarantee' and acknowledge the absence of clinic-based data on success rates. Others refer to the procedure as a 'lottery' but one in which you need to buy a ticket to even have a chance. Like a lottery, it is possible that some women imagine that they will be the exception, the 'winners' in the process. There is also an underlying technological optimism at work here. Wider discourses about bodies (and nature) without limits intersect with the view that all problems will have a technological solution. There appears to be a palpable sense that by the time women become ready to take their eggs out of the bank to be thawed, fertilised, and implanted, the biotech options will have advanced so considerably that there will be no limits to reproductive choice. This is certainly the self-promotional image of the biotech revolution and explains why ovarian tissue freezing is now being promoted as the next big thing. For women who have bio-banked and have undergone the medical steps leading (or not leading) to implantation, much seems to have been invested in the fantasy about the quality of their eggs, alongside a fantasy of being in control (see Bach in this volume).

In the absence of specific research focusing on these fantasies, it may be instructive to return to the female 'high flyer' photographed on the cover of *Bloomberg Business* (2014) with the then-explosive caption 'Freeze your eggs, Free your career'. The woman Brigitte Adams reported at the time to having experienced a wonderful sense of freedom after freezing her eggs. She began a blog advocating the benefits of egg freezing called, *Eggsurance*, and energetically promoted biobanking as a form of empowerment for women. Fast forward to 2017. Adams, at age 44, decided to unfreeze her eggs and start the fertilisation and implantation process so she could begin a family on her own. According to the *Washington Post*:

> Two eggs failed to survive the thawing process. Three more failed to fertilize. That left six embryos, of which five appeared to be abnormal. The last one was implanted in her uterus. On the morning of March 7, she got the devastating news that it, too, had failed.
>
> Adams was not pregnant, and her chances of carrying her genetic child had just dropped to near zero. She remembers screaming like "a wild animal", throwing books, papers, her laptop—and collapsing to the ground. (Cha, 2018)

The failure of assisted reproduction in these circumstances would not be unexpected statistically, given Adams' age. Rather, what stands out is the question she asked herself after all these difficult and distressing procedures did not work. Describing feeling angry, sad, and ashamed, she also asked herself: 'What did I do wrong?' (Cha, 2018). While Adams later renewed her research and advocacy to better inform women to be wary of the hype about social egg freezing, her initial question points to something just as significant. It confirms that the subject being produced as a consumer of egg freezing and addressed by fertility clinics (and in some cases through media representations) is that of an ideal neoliberal individual; an entrepreneurial, self-sufficient, and empowered woman in paid employment, with sole responsibility to manage her own future. In Adams' question she highlights the way this cultural directive to forecast, control, and ensure a particular future for herself is received and translated into individual guilt when things beyond her control go wrong.

A comparable range of difficulties to those of Adams were faced by other women interviewed by Ariana Eunjung Cha for this *Washington Post* article. These included eggs being abnormal, miscarriages, leaks in the straws in which eggs were stored or the case of 18 eggs being destroyed 'while being shipped from one clinic to another'. Even so, once there was a live birth, even if it did not come from the stored and thawed eggs, women described the fertility preservation process as having been a good investment or even the 'best investment ever'. A brief survey of other online articles and posts reveals similar accounts of failures such as damaged eggs and an experience of surprise among women that their eggs seemed to be so quickly 'used up' in attempts at implantation. Nevertheless, contrary to the disillusionment one might expect from these experiences, these women remained optimistic about the technology. They located the problem as a personal failure of time management or inadequate decision-making on their part. If there was regret, it was a regret that they should have started the freezing process earlier or they should have frozen more eggs. As one woman commented after all her eggs failed: 'It was the right choice at the time… The egg freeze gave me peace of mind, and you never know what it can lead to' (Fiona quoted by Fratantoni, 2018).

These stories follow a recognisable pattern. They are invariably accompanied by a photograph of one or more of the women holding a newborn baby conceived through other forms of assisted reproduction or a woman clutching an ultrasound scan photograph of a healthy foetus. At some level, these accounts read like love letters to biotechnology, even though the failures they document along the way are so dramatic. They tap into a pervasive and much broader discourse about technological optimism in the face of biological finitude, either of the human or environmental kind. I am thinking here of proposals to solve fossil-fuelled climate breakdown by space travel or mining asteroids. Of course, there is no direct equivalent between this and reported breakthroughs in fertility preservation, nonetheless, both are based on a similar source and reflect a 'technophilic neoliberalism',[6] an undue attraction to and faith in self-regulating market-based technological innovation as the solution to all social and environmental problems.[7] This confident expectation is evident in the hopeful optimism underlying the stories these women tell, regardless of the success or failure of the egg-freezing technology.

The fact that the procedure continues to grow in popularity at an exponential rate is also an expression of this optimism. Biocapitalist innovation marches ever forward. One biotech company now promises 'robotic cryostorage', hailing it as 'the first fully automated and digital RFID-powered platform that ensures the safety and security of irreplaceable fertility cells' (TMRW, 2020). This same company predicts that on the current rate of growth, in 2025 there will be 2.6 million women storing their eggs or embryos (Kowitt, 2020). On any count, that is a lot of eggs. Given that 'hope' is something that 'consumers' of fertility preservation have already purchased when they begin treatment, it is not surprising that a conviction remains, despite evidence to the contrary, that technology will eventually outwit biological time. Indeed, the biotech industry could be seen to be in the very business of the production and marketisation of hope.

Many commentaries on social egg freezing identify the language of risk management in the advice given to women by the industry, often slanting more towards the possible gains of any treatment than the risks. However, as discussed above, this responsibility is projected onto women themselves as something they need to manage and control at an individual level. Yet, as eggs being damaged in transit or tissue being destroyed by faulty containers show, these more structural, industrial elements are not factored into the promotion of 'the individual insurance against risk' equation. This is especially true regarding the globalisation of biobanking and assisted reproduction. Once a 'happy futurity' or a kind of reproductive immortality is promised, there are contingencies and chains of contingencies that come into play. Far from an individual burden, some of these may have a global dimension. In their introduction to the collection, *The Reproductive Industry: Intimate Experience and Global Processes*, the editors remind us of the fragility of these processes, where natural disasters like the earthquake in Nepal or the tsunami in Fukushima can interrupt power supplies and damage embryos or gametes or disrupt relationships between donors, surrogates, and those 'consuming' these services (Mackie et al., 2019, pp. 2–3). The COVID-19 pandemic is one such emergency that has interrupted cross-border surrogacy arrangements (Mowbray, 2020). No technological utopianism can counter these contingencies, particularly at a time when such events are predicted to become more prevalent.

Egg Freezing and the Maternal

Social egg-freezing is often positioned as a natural extension of the project of mainstream liberal feminism, as I have detailed above. A key element of this dominant variety of feminism is its primary emphasis on fostering women's capacity and opportunity to engage in paid work and to extend women's participation in the public sphere. In the words of Hester Eisenstein, feminism continues to be recognised and understood as paid work for women, and a limited access to power for an elite few. Writing from outside a liberal framework, Eisenstein goes on to define empowerment as 'the incorporation of women into the structures of capitalist power, whether as entrepreneurs or as low-wage workers' (Eisenstein, 2017). This mainstream, liberal or 'marketplace feminism', is not associated with direct acts of protest and resistance. Emancipation as a collective goal is replaced by the injunction to strive towards individual empowerment. Other scholars, myself included, have identified different points at which this liberal project has now been superimposed with neoliberal conceptions of the self, gender, citizenship, and the public/private distinction (Stephens, 2011). According to Catherine Rottenberg, neoliberalism has produced a new feminist subject, one who is entrepreneurial as well as individualised and 'oriented toward optimising her resources through incessant calculation, personal initiative, and innovation' (2020, p. 57). The intersecting points between this version of the empowered feminist and the discourses on social egg freezing are easy to locate. Referencing Angela McRobbie among others, Rottenberg described this feminism as speaking to upwardly mobile, aspirational women, who are 'increasingly being encouraged to invest in themselves and their professions first and to postpone maternity until some later point' (Rottenberg, 2017, p. 332).

The convergence between feminism and neoliberalism has taken on a new complexion that does not so much represent the management of future risks for individual women but rather signifies the promise of 'future individual fulfillment or, more accurately, on careful sequencing of career and maternity and smart (self-)investments in the present to ensure enhanced returns in the future' (Rottenberg, 2017, p. 332). Rottenberg points to how in the popular imagination a new ingredient in the discourse about work/family balance has taken hold. 'Balance' no longer refers to women having to choose between career and family in an either/or scenario, but balancing career and family by cultivating smart, future-oriented investments to ensure that both can be achieved. Timely social egg freezing early in the career of an upwardly mobile middle-class woman would lead to a situation, at some unidentified time in the future, where she could successfully negotiate the pull of a high-flying career with that of a professionalised form of mothering.

Where I would depart from Rottenberg's analysis of this idealised, normative notion of 'balance' for the neoliberal feminist is in the version of motherhood she argues is a key ingredient on the other side of the balance equation. In her view, 'intensive mothering' is envisaged, according to these normative notions, as the path to a putative 'happy balance'. Intensive mothering has come to be a convenient shorthand for a labour intensive, financially costly, and emotionally involving form of mothering. However, I would go as far as questioning whether anything more than a very truncated, amorphous view of motherhood, as purely an abstract potentiality (suspended in time) is evident in the discourse. It is certainly true that when motherhood is represented as a problem for contemporary women, the solution is often presented in the form of better time management. Yet in the 'cost-benefit calculus' identified by

Rottenberg and central to the cultural discourses around social egg freezing as a form of biopreparedness, the risk on the other side of the equation is not so much as missing out on having children but as being forced (by circumstances and biological ageing) to relinquish motherhood as form of individualised fulfilment. In this respect, the tangible, affective, material, and social dimensions of the maternal are completely absent from view. Nowhere is this more evident than in the fertility industry's use of the term 'live baby'. There is a vagueness around actual children, and a silence about the vulnerability of babies and infants, and care and dependency: in short, any sense of human fragility is erased. Clearly, for marketing purposes, this absence is crucial to maintaining technological utopianism and futurity of the biotech imaginary. It is also a part of what Nancy Fraser calls the 'crisis in social reproduction', a 'care crisis' where society's capacity for 'birthing and raising children, caring for friends and family members, maintaining households and broader communities, and sustaining connections more generally' (Fraser, 2014, p. 542) has been depleted.

This crisis in social reproduction is itself reinforced in egg-freezing discourses by presenting the act of getting pregnant and giving birth as an individual woman's responsibility, not as an undertaking also linked to establishing connectedness, or as a way to reproduce kin, families, and heritage in whatever contemporary form. Different theoretical lenses are required to better understand not only the personal impact of a neoliberalising feminism on women undergoing fertility preservation but in order to unearth less obvious public and social impacts of these technologies. Despite the best intentions to the contrary, scholars and commentators find it difficult to move beyond liberal feminist frameworks and ideologies of choice as empowerment when debating this issue. As mentioned earlier, psychoanalytic and psychosocial approaches may better register the fantasies women have about their stored eggs, or the irreconcilable desires and imaginings that may underpin the urge for a genetically related child. Such research could also investigate the important distinction made by Damien Riggs (2019, p. 70) between desiring to preserve one's fertility and desiring to have a baby, as one may not necessarily follow or be related to the other. Importantly, a maternal feminist frame of interpretation is needed to shift the terms of the debate away from choice and empowerment towards questions of what else may be at stake at the level of social (rather than individual) reproduction. To return to Fraser's notion of crisis, social reproduction may be what is sacrificed in our cultural and technological optimism about reproductive futurity.

Finally, maternal feminism is the most effective counter to neoliberal feminism because it focuses on connectedness, human vulnerability, care, and protection and the formation and preservation of deeply embedded social ties, in short, social reproduction. It is a framework founded on the premise of recognising the *public* importance of mothering and the care of children and acknowledging that the ideals and ethics associated with maternal care have important social and political value. Fertility preservation, viewed through this lens, may gesture towards a more complex picture of social egg freezing and the wider cultural impact of the production of a 'potentially maternal' subject. The role played by pervasive anxieties about motherhood in convincing women to freeze their eggs needs further investigation as well as overcoming the structural barriers contemporary mothers face, not only in the workplace. In this cultural moment, motherhood as self-actualisation is deemed unproblematic. By contrast, motherhood conceived of as involving care, sacrifice, human dependency, and being connected to the needs of children is seen as too culturally laden, too weighed down by conflicting associations to be acceptable in neoliberal times. Does the 'hype' associated with social egg freezing, both of the negative and positive kind, reinforce these dominant cultural anxieties around the maternal? If so, certain versions of the maternal are also being put in the freezer alongside all those eggs. This impacts our understanding of care and human fragility together with social and intimate relations in ways that are yet to be fully configured and understood.

Notes

1. Some radical feminist critiques of the fertility industry are based on the premise that better knowledge about the medical harms inflicted by reproductive technology or the exploitation of women in human surrogacy arrangements would release people from their desire to have a genetically related child (e.g., Klein, 2017, p. 176).

2. For an outstanding example of this approach, see Lois Tonkin (2017).
3. See Chap. 4 in this volume.
4. 'Frozen egg use in IVF doubled in two years in Victoria' (Victorian Assisted Reproductive Treatment Authority, 2018).
5. To view a recent reproduction of this cover of *Bloomberg Business*, see Genetic Literacy Project (2020).
6. A term used by Atus Mariqueo-Russell and Rupert Read (2019).
7. For a discussion of this utopianism, see Boris Frankel (2018).

References

Adora Fertility. (2020). Egg freezing. Retrieved February 19, 2020, from https://www.adorafertility.com.au/egg-freezing/

Alteri, A., Pisaturo, V., Nogueira, D., & D'Angelo, A. (2019). Elective egg freezing without medical indications. *Acta Obstetricia et Gynecologica Scandinavica, 98*(5), 647–652.

Argyle, C. E., Harper, J. C., & Davies, M. C. (2016). Oocyte cryopreservation: Where are we now? *Human Reproduction Update, 2*(4), 441–442.

Bach, A. S., & Kroløkke, C. (2020). Hope and happy futurity in the cryotank: Biomedical imaginaries of ovarian tissue freezing. *Science as Culture, 29*(3), 425–449.

Baldwin, K. (2019). The biomedicalisation of reproductive ageing: Reproductive citizenship and the gendering of fertility risk. *Health, Risk & Society, 21*(5–6), 268–283.

Baldwin, K., & Culley, L. (2020). Women's experiences of social egg freezing: Perceptions of success, risk and 'Going It Alone'. *Human Fertility, 23*(3), 186–192.

Bühler, N. (2015). Imagining the future of motherhood: Medically assisted extension of fertility and the production of genealogical continuity. *Sociologus, 65*(1), 79–100.

Bühler, N. (2021). *When reproduction meets ageing: The science and medicine of the fertility decline*. Emerald Publishing Limited. https://doi-org.helicon.vuw.ac.nz/10.1108/978-1-83909-746-120211001

Cha, A. E. (2018, January 27). The struggle to conceive with frozen eggs. *The Washington Post*. Retrieved January 28, 2020, from https://www.washingtonpost.com/news/national/wp/2018/01/27/feature/she-championed-the-idea-that-freezing-your-eggs-would-free-your-career-but-things-didnt-quite-work-out/

Chodorow, N. (2003). "Too Late": Ambivalence about motherhood, choice and time. *Journal of the American Psychoanalytic Association, 51*(4), 1181–1198.

Cooper, M. (2008). *Life as surplus: Biotechnology & capitalism in the neoliberal era*. University of Washington Press.

Eisenstein, H. (2017). Hegemonic feminism, neoliberalism and womenomics: 'Empowerment' instead of liberation? *New Formations, 91*, 35–49.

Eventbrite. (2020). Eggs in the city. Retrieved October 22, 2020, from https://www.eventbrite.com/e/eggs-and-the-city-tickets-80303611263

Frankel, B. (2018). *Fictions of sustainability: The politics of growth and post-capitalist futures*. Greenmeadows Press.

Fraser, N. (2014). Can society be commodities all the way down? Post-Polanyian reflections on capitalist crisis. *Economy and Society, 43*(4), 541–558.

Fratantoni, M. (2018, November 3). Fiona's story. *The New Daily*. Retrieved December 15, 2019, from https://thenewdaily.com.au/life/wellbeing/2018/11/03/fionas-story-chose-freeze-eggs/

Fyfe, M., & Davies, J.-A. (2015). Motherhood on ice. *The Sunday Morning Herald*. Retrieved October 22, 2020, from https://www.smh.com.au/lifestyle/motherhood-on-ice-20150603-ghfwga.html

Genetic Literacy Project. (2020). Egg freezing: Smart career move? Retrieved February 23, 2020, from https://geneticliteracyproject.org/2014/10/21/egg-freezing-a-smart-career-move/

Harvey, D. (2007). *A brief history of neoliberalism*. Oxford University Press.

Harwood, K. (2009). Egg freezing: A breakthrough for reproductive autonomy? *Bioethics, 2*(1), 39–46.

Kato, K. (2016). Vitrification of embryos and oocytes for fertility preservation in cancer patients. *Reproductive Medical Biology, 15*(4), 227–233.

Klein, R. (2017). *Surrogacy: A human rights violation*. Spinifex Press.

Kowitt, B. (2020, January 22). Fertility Inc.: Inside the big business of baby-making. *Fortune*. Retrieved October 21, 2020, from https://fortune.com/longform/fertility-business-femtech-investing-ivf/

Lockwood, G., & Fraser, B. C. J. M. (2018). Social egg freezing: Who chooses and who uses? *Reproductive Biomedicine Online, 37*(4), 383–384.

Mackie, V., Marks, N. J., & Ferber, S. (Eds.). (2019). *The reproductive industry: Intimate experiences, global processes*. Rowman & Littlefield.

Mariqueo-Russell, A., & Read, R. (2019). Fully automated luxury barbarism. *Radical Philosophy, 2.06*(2), 108–110.

Mowbray, N. (2020, June 14). Family planning: How COVID-19 has placed huge strains on all stages of surrogacy. *The Guardian*. Retrieved October 19, 2020, from https://www.theguardian.com/lifeandstyle/2020/jun/14/family-planning-how-covid-19-has-affected-all-steps-of-surrogacy

Petropanagos, A., Cattapan, A., Baylis, F., & Leader, A. (2015). Social egg freezing: Risk, benefits and other considerations. *Canadian Medical Association Journal, 187*(9), 666–669.

Pritchard, N., Kirkman, M., Hammarberg, K., McBain, J., Agresta, F., Bayly, C., Hickey, M., Peate, M., & Fisher, J. (2018). Characteristics and circumstances of women in Australia who cryopreserved their oocytes for nonmedical indications. *Journal of Reproductive and Infant Psychology, 35*(2), 108–118.

Radin, J. (2015). Planned hindsight. *Journal of Cultural Economy, 8*(3), 361–378.

Riggs, D. W. (2019). An examination of 'Just in Case' arguments as they are applied to fertility preservation for transgender people. In V. Mackie, N. J. Marks, & S. Ferber (Eds.), *The reproductive industry* (pp. 70–71). Rowman & Littlefield.

Riggs, D., & Bartholomaeus, C. (2018). Fertility preservation decision making amongst Australian transgender and non-binary adults. *Reproductive Health, 15*(1), 1–10.

Rosewarne, L. (2014, October 15). Is corporate egg freezing such a rotten idea? *ABC News*. Retrieved February 22, 2020, from https://www.abc.net.au/news/2014-10-15/rosewarne-is-corporate-egg-freezing-such-a-rotten-idea/5815686

Rottenberg, C. (2017). Neoliberal feminism and the future of human capital. *Signs: Journal of Women in Culture and Society, 42*(2), 329–348.

Rottenberg, C. A. (2020). *The rise of neoliberal feminism*. Oxford University Press.

Sample, I. (2019, August 4). New medical procedure could delay menopause by 20 years. *The Guardian*. Retrieved February 16, 2020, from https://www.theguardian.com/science/2019/aug/04/medical-procedure-delay-menopause/

Shady Grove Fertility. (2020). 5 reasons to freeze your eggs. Retrieved January 22, 2021, from https://www.shadygrovefertility.com/blog/treatments-and-success/5-egg-freezing-benefits-and-why-you-should-consider-freezing/

Shaw, R. M. (2017). Bioethics beyond altruism. In R. M. Shaw (Ed.), *Bioethics beyond altruism: Donating & transforming human biological materials* (pp. 3–35). Palgrave Macmillan.

Stephens, J. (2011). *Confronting postmaternal thinking: Feminism, memory and care*. Columbia University Press.

Stoop, D., van der Veen, F., Deneyer, M., Nekkebroeck, J., & Tournaye, H. (2014). Oocyte banking for anticipated gamete exhaustion (AGE) is a preventive intervention, neither social nor nonmedical. *Reproductive BioMedicine Online, 28*, 548–551.

TMRW. (2020). Software guided cryospecimen management. Retrieved October 20, 2020, from https://www.tmrw.org/

Tonkin, L. (2017). 'A sense of myself as a mother': An exploration of maternal fantasies in the experience of 'circumstantial childlessness'. *Studies in the Maternal, 9*(1). https://doi.org/10.16995/sim.244

Van de Wiel, L. (2014). For whom the clock ticks: Reproductive ageing and egg freezing in Dutch and British news media. *Studies in the Maternal, 6*(1), 1–28. https://doi.org/10.16995/sim.4

Van de Wiel, L. (2015). Frozen in anticipation: Eggs for later. *Women's Studies International Forum, 53*, 119–128. https://doi.org/10.1016/j.wsif.2014.10.019

Victorian Assisted Reproductive Treatment Authority. (2018, September). Frozen egg use in IVF doubled in two years in Victoria. Retrieved February 23, 2020, from https://www.varta.org.au/resources/news-and-blogs/frozen-egg-use-ivf-doubled-two-years-victoria

Weinbaum, A. E. (2019). *The afterlife of reproductive slavery: Biocapitalism and Black feminism's philosophy of history*. Duke University Press.

Part II

Rights

6

Reproduction and Beyond: Imaginaries of Uterus Transplantation in the Light of Embodied Histories of Living Life Without a Uterus

Lisa Guntram

Introduction

Charlotte[1] was in her mid-teens when she learned that she did not have a uterus. When we met, several years had passed since then but she had only told a few persons about her condition. She described how she often felt that people did not understand what she, and other women in similar situations, had been through.

> People don't have a clue. […][2] [In the media] there could be greater emphasis on the psychological aspects and on how bad you feel. All of those who commit suicide because of this condition, it happens. It's proven, too. That it leads to that. […] To emphasise that a bit more, to discuss that more. To make people see what it is really about. That it is not simply about getting a uterus.

L. Guntram (✉)
Linköping University, Linköping, Sweden
e-mail: lisa.guntram@liu.se

© The Author(s), under exclusive license to Springer Nature Singapore Pte Ltd. 2022
R. M. Shaw (ed.), *Reproductive Citizenship*, Health, Technology and Society,
https://doi.org/10.1007/978-981-16-9451-6_6

Approximately 1 in 500 women of reproductive age—about 2000 women in Sweden, and 200,000 women in Europe—live, like Charlotte, with infertility caused either by congenital or acquired absence of the uterus, or by a 'malfunction' of the uterus (Kvarnström et al., 2017). These conditions are also known as 'uterine-factor infertility' (UFI) and prevent the person from carrying a pregnancy (Brännström et al., 2014). During the past two decades research in uterus transplantation as a means to cure UFI has developed considerably. At the forefront, Swedish researchers have, since the late 1990s, investigated the prospect of uterus transplantation in which a living donor gives their uterus to a recipient and initiated the first systematic uterus transplantation trial (Brännström, 2015, 2019; Brännström et al., 2020).

In the medical literature concerning uterus transplantation (which is combined with IVF and has been given the abbreviation 'UTx-IVF' in this chapter), women are commonly presented as requesting the procedure because it can deliver genetic and gestational parenthood, which alternatives such as surrogacy and adoption cannot (Arora & Blake, 2015; Testa et al., 2017b). The procedure is thus commonly not considered fully successful until it has resulted in the birth of a healthy child (see, e.g., Brännström et al., 2018; Järvholm et al., 2015). In contrast, social scientists and feminist scholars have often argued that assisted reproductive technologies do not simply treat infertility (Becker, 2000; Franklin, 1997, 2013; Lie & Lykke, 2016). Charis Thompson (2005, p. 8), for example, posits that these technologies produce 'parents, children and everything that is needed for their recognition as such'. However, little is known about the motivations and desires of women with UFI when it comes to pursuing or not pursuing UTx-IVF (Farrell et al., 2018; Horsburgh, 2017; Guntram, 2018).

In this chapter, I examine the accounts of ten Swedish women living with a congenital absence of the uterus and the entire or parts of the vagina.[3] The condition, also known as Mayer-Rokitansky-Küster-Hauser syndrome (MRKHS), is found in about 1/4000 to 1/10,000 women. Affected individuals cannot menstruate or conceive because of the absent uterus and may, depending on the size of the vagina, experience difficulties to perform vaginal penetration (ACOG, 2002). In terms of treatment the absent or small vagina is typically enlarged surgically, after puberty, if the woman so wishes. As MRKHS does not affect the ovaries, women with the condition may have eggs retrieved to be used in surrogate pregnancy arrangements or pregnancy following UT-IVF.

The ten women I interviewed had all considered a future with a uterus through UTx. In this chapter, I specifically explore what they imagine this innovation—in its particular context, applicable to individuals with specific hopes and histories—can enable them to *do* or *become*. With this as my starting point, I am interested in the imagined futures that arise from the prospect of UTx-IVF, and how these comingle with the accounts of women living without a uterus. More specifically, I aim to demonstrate how social imaginaries—collectively formed meanings and affects that shape how we experience the social world and our bodies (Hudson, 2020)—come to materially shape bodies, life trajectories, and desires, and consequently also approaches to UTx-IVF. I do this by examining how the study interviewees themselves make sense of UTx-IVF in the light of their past experiences. To conclude, I discuss the significance of the histories of those involved, and of shared social imaginaries of reproductive liberty, medical prospects, and female embodiment, when seeking to understand women's motivations for pursuing medically assisted human reproduction (AHR). I argue that such considerations must be taken into account in policy discussions about the development of UTx-IVF.

Medical Materialities of UTx-IVF and the Routinization of Medical AHR

An examination of the social imaginaries enacted in the prospect and process of UTx-IVF requires consideration of both the specificities of the procedure and the normative forces of AHR. Swedish research on uterus transplantation, in keeping with all reported attempts around the world, combines uterus transplantation with IVF. In practice, this means that following extensive assessment of the donor, the recipient, and the recipient's partner (including both psychological tests and physical matching), ovarian eggs are harvested from the recipient and subsequently fertilized *in vitro*. The cervix, the uterus, and two major blood vessels are removed from the donor and connected to the vagina of the recipient. The recipient is given immunosuppressive treatment to avoid rejection. About a year later, if she has started menstruating and the organ is stable, embryo transfers are initiated. If she becomes pregnant, the child is delivered by caesarean section. The recipient may then attempt a second pregnancy if she is assessed to be medically fit (Brännström et al. 2018; Brännström 2019). It should also be mentioned that a transplanted uterus is not intended to remain in place for life. It may be used for two pregnancies, after which, if it does not endanger the health of the recipient, it is removed. This is to minimize the time that is necessary to take immunosuppressive drugs (Brännström et al., 2015). As of July 2019, 14 children had been born as a result of UTx-IVF around the world, 9 of them by women participating in the Swedish research project (Brännström et al., 2019).

The medical risks for recipients of UTx-IVF are associated with several aspects of the procedure: the three major abdominal surgeries it involves (UTx itself; the caesarean section to deliver the baby; the hysterectomy after the delivery or, in some cases, after a second delivery), immunosuppressive therapy, ovarian stimulation for IVF (which may lead to, e.g., ovarian hyperstimulation syndrome), and oöcyte (ovarian egg) pick-up (which may lead to, e.g., ovarian bleeding). If a pregnancy is achieved, there are risks of preeclampsia and premature birth (Testa et al., 2017a). In addition, organ transplantation in general—not only UTx—involves embodied, relational, and emotional contingencies and difficulties (see, e.g., Gunnarson, 2016; Papachristou et al., 2009; Scheper-Hughes, 2007). The same is true for the IVF procedure that is a part of UTx-IVF (see, e.g., Franklin, 1997; Peters et al., 2007; Thompson, 2005; Throsby, 2004).

As UTx-IVF has emerged as a viable procedure, the ethical discussions about the procedure have included a risk-benefit analysis (Arora & Blake, 2015; Catsanos et al., 2013; Testa et al., 2017a), questions of priority-setting and access to treatment (Jones et al., 2019; Sandman, 2018; Wilkinson & Williams, 2016), and regulatory difficulties and reproductive rights (Alghrani, 2018a, 2018b; Horsburgh, 2017). In addition it has been pointed out that further examinations, taking into consideration the prospective recipient's lived experiences, motivations, and perceptions of the benefits are necessary (Horsburgh, 2017), as well as analyses that consider the utility, risks, and benefits of UTx-IVF to be culturally embedded (Farrell et al., 2020). For example, Richards et al. (2019) showed that women who seek to undergo UTx-IVF describe the procedure as a way to gain or regain reproductive autonomy and to strengthen family intimacy (see also Saso et al., 2016). The need for studies that address social expectations and norms concerning parenthood and gender has also been underlined (Allyse, 2018; Farrell et al., 2018; Horsburgh, 2017; Mabel et al., 2018; Mertes & Assche, 2018). Specifically, concerns have been raised that some of the proposed criteria for taking part in research on UTx-IVF are overly regulatory of women's bodies (Mabel et al., 2018) and that women who seek UTx-IVF must meet a 'much more stringent list of motivations to justify the 'risk' to their bodies of uterine transplant' than men seeking penile transplantation (Allyse 2018, p. 35). For example, the list of motivations for women who wish to undergo UTx-IVF typically includes a desire to experience pregnancy and to become a mother, and not simply a wish to restore bodily integrity (Bayefsky & Berkman, 2016).

The normative dimensions of medical practices and treatments have been a core concern in much feminist scholarship, which has demonstrated the regulatory and at the same time disruptive force of medical AHR. While some scholars have underlined the material-discursive entanglement of the 'social' and 'technological' in AHR (Franklin & McKinnon, 2001; Murphy, 2012; Thompson, 2005), others have made explicit the experiences and negotiations of such procedures (Exley & Letherby, 2001; Letherby, 2002; Mamo, 2007; Nordqvist, 2008; Peters et al., 2007; Throsby, 2004). Feminist examinations of medical AHR also demonstrate women's experiences of restricted choice and the routinization of medical interventions. For instance, it was suggested in the 1980s that the social stigma of infertility and of not becoming a mother caused women to feel that they had little choice but to use medical interventions (Ulrich & Weatherall, 2000). More recently, Sarah Franklin has argued that 'conception *in vitro* is now a normal fact of life' (Franklin, 2013, p. 1). Franklin's observation is that assisted reproduction has become naturalized and normalized to the point that it is taken for granted as a routine and integral part of healthcare (Franklin, 2013, pp. 4, 6) and of family-building processes (Gammeltoft & Wahlberg, 2014). In light of this, medical decisions undertaken by individuals, including those about whether to pursue medical AHR, can be understood not only as a matter of pathology but also as a matter of desires shaped by market culture, social aspirations, and the fertility industry (Kroløkke & Kotsi, 2019; Whittaker, 2011).

Methodological and Analytic Framework

The overarching project, which includes the findings presented in this chapter, examines UTx-IVF from the perspectives of potential participants and of the healthcare professionals involved. The qualitative design of the project was approved by the Regional Board for Ethical Review, Linköping (EPN, 2014). Ten women participated in the study. They had all discovered in their teens that they did not have a uterus, and had all considered, to different extents, UTx-IVF. Some had been accepted into the Swedish research project on UTx-IVF and had undergone transplantation; others had not been accepted for the project; others still had opted for surrogacy abroad or adoption. I interviewed six of the ten participants once, had one follow-up interview with one participant, and had two follow-up interviews with three participants (who were in the process of transplantation). I used a semi-structured interview guide, which was developed as the project evolved and I strived to maintain a reflexive and sensitive approach to how my position as a researcher affected my relationship with the interviewees (c.f. Burns, 2003; Ellingson, 1998). In the interviews, I also endeavoured to let the interviewees know about my research interests, while avoiding the claim that I could fully grasp the complexities and nuances of their embodied experiences and histories.

I used the analytic software Atlas.ti to transcribe the audio recordings, and to conduct a general thematic coding. During this process, I identified two core analytic questions around which I organized the analysis: 'What is UTx-IVF doing for the interviewees?' and 'What does UTx-IVF make them do?'[4] After mapping the material according to these questions, I employed conceptualizations of social imaginaries and bodily imaginaries (Dawney, 2011; Hudson, 2020; Lennon, 2004), as an analytic lens when re-reading the interviewees' accounts. Doing so, I considered social imaginaries to be both collectively formed and personally embodied (Hudson, 2020; Lennon, 2004). This makes it possible to understand social imaginaries not only as discursive entities, but as corporeally produced and reproduced through technologies and practices. In other words, social imaginaries are *embodied imaginaries* (Hudson, 2020, p. 348). Gendered ideas, for example, produce 'affective relations which are central to the production of subjects,' through which 'bodies are constrained and enabled' (Dawney, 2011, p. 541). In her work on imaginary embodiment Kathleen Lennon (2004) focuses specifically on *bodily imaginaries*, and considers bodies to be formed as they are invested with affect. This leads her to suggest that bodily imaginaries constitute our bodily identities. Sexual difference, argues Lennon (2004, p. 116), is:

> constituted out of the imaginary investments in different body parts, the salience we attach to them. The body, as it features in these accounts, is a body experienced not only cognitively but also affectively. That is, we do not only categorise the bodies of ourselves and others, we imagine them, and the way we imagine them structures the formation of our desires.

Further, Nicky Hudson (2020, p. 348) suggests that accounts of medical AHR involve, for instance, 'deliberations, thoughts, feelings, ambivalences and expectations about treatment, alongside actions more traditionally conceived of as material practices—that is, those relating to embodied experiences of drug regimens or clinical encounters.' These conceptualizations of imaginaries allowed me not only to analyse experiences of UTx-IVF, but also to examine how the interviewees' accounts of the motivations, desires, expectations, and promises attached to UTx-IVF were connected to larger social imaginaries of female embodiment, and to their specific history of living without a uterus. The accounts from the interviewees and the emergence of UTx-IVF could thus be understood as formed by the collective meanings and affects that shape the way we experience the social world and our bodies (Dawney, 2011; Gatens, 1996; Hudson, 2020).

Shaping Desires, Bodies, and Expectations. The Entanglements of Social Imaginaries

There are many ways to describe the experience of discovering as a young woman that you do not have a uterus. The core may be a feeling of distortion. Something you had assumed about your body, about what was inside it and what you could do with it, is suddenly shattered (see Guntram, 2013; Zeiler & Guntram, 2014). The interviewees in the present study described the disruptive force of being unable to follow the reproductive trajectory they had expected, similar to that experienced by people with other forms of infertility (Exley & Letherby, 2001; Sandelowski et al., 1991; Throsby, 2004). However, since the absence of a uterus is commonly detected in puberty (Herlin et al., 2016), the women I interviewed had faced the consequences of their condition for their reproductive trajectory much earlier than women who discover infertility when trying to become pregnant (Guntram, 2018). For most of them, the realization that they would not be able to have children as a consequence of the diagnosis was devastating (see also Holt & Slade, 2003; Guntram, 2018; Richards et al., 2019; Zeiler & Guntram, 2014). However, not all interviewees described the desire to undergo pregnancy as paramount. They tended to focus on their wish for a child and several had considered, and some had pursued, different routes to parenthood such as surrogacy and adoption.

In what follows I do not mean to disregard that many women, including some of the interviewees in my study, view UTx-IVF mainly as a response to experience pregnancy (Guntram, 2018) or the desire to have a child in any way possible. At the same time, other emotions in my interviewees' accounts seemed to be entangled with the desire to have a child, as too were past experiences and expectations. Below, I explore such themes that do not directly concern parenthood and reproduction, but other dimensions of UTx-IVF.

The Same Options as Everyone Else: Imaginaries of Reproductive Trajectories

As I have mentioned above, success in UTx-IVF is often defined in the medical literature as the birth of a healthy child (Brännström et al., 2018). While it is true that a central goal among the interviewees was to have a child, other concerns came to light as they reflected on why they had considered UTx-IVF.

When we first met, Sophia had only recently started to consider UTx-IVF. She had thought a great deal about what it would mean to share the experience of having a child with her partner, and how she felt about having or not having a genetic link to that child. She now felt that the biological aspect might not matter that much to her, but added that it felt difficult to not have any alternatives or make her own choices:

Sophia: You simply want the same options, you know, like everyone else.
L: You return to this about options, that the important part is to have options?
Sophia: Yes, maybe.
L: You said it a couple of times, that it is about not having the same options, to not even have the possibility to try […]. Is that something you've noticed, or is it just me?
Sophia: Yes, yes but, well, there may be something in that. Absolutely. That you would want all alternatives open, you know. To see that you, that you choose, you know, how you would like your future to be, in some way. And then there is still a route which, so far, is "locked" and closed. So that's true, it's probably a lot about the options.

This conversation demonstrates how the specificities of Sophia's embodiment made her aware that certain routes to parenthood were not available to her, and how she had aligned her desires for an imagined future accordingly. Sophia's narrative made her ideas about the reproductive possibilities that other women have especially conspicuous; she simply wanted the same options as everyone else. When Olivia described her experience of UTx-IVF, she also related to the assumed experiences of others. She and her partner had initially applied to adopt. 'We were positive at first,' Olivia explained, but described how they had then become disillusioned with the adoption process. 'It was so formidable and unwieldy.' They questioned the need for mandatory classes, the review of their economic situation, and the overall scrutiny of their life. They did not like being steered by someone else, Olivia added, nor that 'someone else would decide our destiny.' For Olivia, who at the time of our interview had undergone a transplantation, it had never been that important to acquire a uterus. 'I had not been longing for a uterus before the transplantation,' she explained. Rather, UTx-IVF offered an opportunity for her to reach the baseline of everyone else. It was a long process, but when she had finally undergone the transplant and was waiting for a year to pass before performing the first embryo transfer, she and her partner concluded that:

> Now we are in the same place as everyone else. You know, now we've got… Now we've got a chance! And that was cool! I felt.

I find the emphasis that Olivia placed on 'everyone else' to be particularly significant, as it indicates the idea of a 'baseline'; a point at which others—those with a uterus—are located. After reaching this baseline by acquiring a uterus, you have the same chance as anyone else to become a parent—with the help of medical AHR. In these accounts, the potentiality of UTx-IVF—in the sense of a set of possibilities, or the 'flip side of risk' of the procedure (Taussig et al., 2013)—provides women with UFI with the same options as everyone else. As such, its potentiality includes the possibility of imagining one's future reproductive trajectory as 'open to choice' (Taussig et al., 2013, p. 4), aligning it with the trajectories that are imagined to be shared by others.

The interviewees also spoke about their desire to determine their own reproductive trajectories. This was exemplified in Amelia and Jessica's separate discussions about what they missed and what they wished for. Amelia had considered both UTx-IVF and surrogacy. Just like Olivia, she said that she could manage without the other things—such as menstruation and pregnancy—that come with a uterus: 'The only thing I miss is that I cannot choose myself when I want to start a family.' Instead, she explained, 'It is up to everyone else.' Jessica also emphasized the importance of being able to make her own reproductive decisions. This had become even more significant for her, as she was currently both undergoing medical assessment for UTx-IVF and had earlier started to set up a surrogacy arrangement abroad.[5]

> I would be totally devastated if I could not... I would rather want to make the decision myself. That is, to not use my own eggs for surrogacy or that I'd say, "No I don't want to do this [a transplant] in Gothenburg" if it would come to that. Rather than someone telling me, "This is not possible" because then I don't know... I don't know how I would react.

I have previously shown that UTx-IVF can be construed as enabling a 'corporeal connectedness' with a 'child-to-be,' and with others close to the recipient (Guntram, 2018). However, the interviewees in the present study did not typically express a desire for the experience of pregnancy as such, but rather a desire to be able to choose for themselves. The accounts of both Amelia and Jessica underline this wish to be able to choose among options. As Jessica indicated, this was also the case when UTx-IVF as a route to parenthood was close to becoming reality. Hence, although UTx-IVF may not be preferable to surrogacy in any specific respect, Jessica underlined the importance of feeling that she was the one turning options down, rather than being turned down. As such, her account illustrates the significance of acquiring control of one's reproductive trajectory and ultimately over one's embodied future. The emphasis on options and choice seen in the above accounts brings out dimensions of UTx-IVF that go beyond enabling genetic parenthood and pregnancy, which in ethical discussions have been posited as the motivation for women to seek UTx-IVF (see Richards et al., 2019). Taking the situatedness of social imaginaries into consideration, it is important for me to also read these accounts in the light of discourses of reproductive liberty, in which medical AHR has been praised for increasing 'women's reproductive rights by providing new choices and options' (Beckman & Harvey, 2005, p. 8) and for empowering women. Such understandings of choice become particularly significant in the case of UTx-IVF, as the procedure has sometimes been presented as a way to treat the 'only major cause of female infertility that remains untreatable' (Johannesson et al., 2015), as 'the only treatment for absolute UFI' (Richards et al., 2019, p. 24), and as providing an additional route by which women with UFI can become parents (Brännström et al., 2018). When understood in this way UTx-IVF becomes not simply a response to pathology but an opportunity to grant reproductive liberty that allows women to make choices like 'everyone else' (see Krøløkke & Kotsi, 2019; Whittaker, 2011).

Kathleen Lennon suggests (2004, p. 116) that bodies are not only categorized but also imagined and that the ways in which we imagine them shape our desires. Different imaginaries are, moreover, 'tied up with different ways of responding to and acting in relation to our environment,' and are, as such, 'constitutive of, not merely reflective of, the forms of sociability in which we live' (Gatens & Lloyd, 1999, p. 143). At the same time, images of the body are reflected back to us from our social environment—such as in intimate family relationships and public writings—and are incorporated into how we experience our bodies (Lennon, 2004). In the light of this framing the emphasis placed by my interviewees on options and choice can be understood to tie into a broader social imaginary of the promises of medical AHR. Such an imaginary makes it clear how these technologies are coming into routine use and becoming more readily available, making them an integrated aspect of life around the globe (Franklin, 2013; Gammeltoft & Wahlberg, 2014) and supporting the idea that medicine has (unlimited) possibilities to treat infertility. At the same time, the introduction of new technologies in the field of AHRs highlights differences in legal frameworks and accessibility, differences which affect reproductive citizenship in contexts where reproductive capacity is increasingly understood as 'a central marker to citizenship' (Riggs & Due, 2013, p. 957).

However, while shared social imaginaries should be recognized and problematized, individual differences in how we reflect and incorporate social imaginaries must also be taken into account (Lennon, 2004). To simply consider the interviewees' desire for options and choice as an attempt to align themselves with a social imaginary of reproductive liberty would be to fail to consider the complexities and specificities of this particular technology and it risks portraying the women who use it as simply dazzled by technological advancement (see also Ulrich & Weatherall, 2000). I provide below more details from the interviewees' accounts of their histories that are particularly salient to develop a more nuanced understanding what UTx-IVF may enable subjects to *do* or *become*.

Making Sense of UTx-IVF in the Light of Life Without a Uterus: Imaginaries of Female Embodiment

In our conversations about UTx-IVF, the interviewees returned to their experience of living without a uterus. Most of them explicitly mentioned the acquisition of a uterus only in terms of treating infertility, yet they remarked on what the absence of a uterus had meant and done for them in the past and the present. Many described their confusion as teenagers, and explained how they had dealt with the many questions concerning identity and reproductive trajectory that the condition had given rise to (see Guntram, 2013, 2018). When Olivia, for example, was told at the age of 16 that she did not have a uterus and that she would not be able to 'have children of her own,' she felt it was definite. 'I cried and screamed,' Olivia said, 'and I felt very devastated because I was different—it felt so unfair.' After a while, she gradually began to process it. 'I started to emphasise other qualities instead,' Olivia explained:

> So I didn't bury myself in this, but I slowly and surely added layers of other positive stuff, on top. But it was always kind of there at the core during all these years, you could say.

Even though the condition was always present, Olivia emphasized how she dealt with it by focusing on other parts of her life. Several interviewees described such a redirection of focus. Jennifer, for example, talked about how she had 'compensated' for her condition. After finding out, Jennifer said, 'I felt like a freak,' and she heard the word used by the midwife who had informed her—'malformed'—repeated in her head. Jennifer felt 'disgusted' by herself. She 'didn't feel like a woman—in terms of what a woman should be and look like.' Jennifer explained how this led to her compensating in other ways; 'I had my breasts done when I was 19 in order to, well, compensate in a way for the femininity that had, you know, been stolen from me, in some way.' 'I sought confirmation from men,' Jennifer added, 'and had quite an extravagant way of living in my early 20s in terms of clothes, makeup, hair, nails.'

Jennifer did not have any regrets about her past, which had been lots of fun, but she felt that she was not compensating in this explicit way anymore. At the same time, she explained, she found there was 'still a tendency'. 'It has become a part of my personality, in a way.'

> Now I have known half of my life that I don't have a uterus and all that. It made a deep mark and I'm still struggling a bit, kind of. To... yes... stop compensating. There is no need, I'm married and satisfied and so on... but it... it's kind of so deeply rooted in you, in a way.

And, she added, 'It's society's expectations, that's all it is, I believe.' If society had been more accepting, Jennifer explained, and had not put so much emphasis on femininity and masculinity, she would have been able to come to terms with her condition more readily. Jennifer's account shows, I suggest, how shared social imaginaries of female embodiment came to shape her experiences of living and dealing with the absence of a uterus. Her condition contributed to a process in which she materially altered her body to make it fit a shared social imaginary. While she attained another kind of acceptance, her experience of managing her condition in this way was central to her experience of embodiment. This experience was, in her account, informed in turn by 'society's expectations,' and alternative expectations could have led her to shaping her body differently—that is, other responses and social imaginaries would have made different forms of agency possible (Lennon, 2004, p. 117).

The accounts of Charlotte, Mia, and Emily also showed how previous experiences shaped the ways in which they made sense of their expectations for UTx-IVF. Charlotte explained that she regretted the kind of surgery she had undergone to enlarge her vagina, and that she therefore was unsure whether to try for transplantation:

> I don't know if I want to make the same mistake again, since this is rather new. I don't want to be one of the first ones, you know.

In the same line of reasoning, Mia said, laughing at first, that:

> I'm something of a coward when it comes to these things. I am afraid that it won't turn out well. That I won't feel well. Because I've had this surgery on the vagina and that wasn't easy. So I can imagine that this is even worse because I've read that you get lots of medication and treatment.

Both Charlotte and Mia described how their trajectories of medical interventions had come to shape how they knew themselves and their approach to UTx-IVF. In this way, their medical histories influenced the ways in which they construed their medical futures. Even more explicitly, Emily related her experiences of living without a uterus to her expectations of a life with a uterus, provided through transplantation. Before, Emily said, she had absolutely wanted to carry a child herself. Now, in light of her efforts setting up a surrogacy arrangement abroad, this aspect did not matter as much. Even so, Emily said that if she were offered a uterus transplant, she would accept without hesitation. Acquiring a uterus would, in her mind, mean something more than gaining more options: 'It's all that about my femininity. Or to feel complete. Because now I don't feel complete.'

When Emily found out at age 14 that she did not have a uterus or vagina, she wished that there would come a time when she didn't have to think about it, if only for a day. She would be so happy, she believed, if that happened.

> And you thought that you'd grow up, get more sensible, that you'd grow into it and that you'd become braver. And that that day would come. But it hasn't come. […] I understand that I don't have a uterus, but… I have learned to live with it, but I've never learned to accept it.

Her thoughts, she said, leave her sleepless at night and make her energetically clean her apartment.

> I think that if I could have a transplantation and get the opportunity to go through all that I'd get another kind of peace inside. In some way, I don't think it's about me liking a clinically clean home or that. It is because I have control over it, and it keeps me busy. 'Cause as soon as I sit down and think too much, life gets so hard. Then, then I panic.

The directness with which Emily explained what a uterus would mean to her differs from the accounts of the other interviewees. In this sense, her account does not illustrate a general pattern in the material, nor may it represent the experience of most women affected by the condition. Rather, it underlines the multifaceted ways that women make sense of UTx-IVF in light of their condition as it shows how one may experience embodiment and the perceived promises of UTx-IVF in relation to one's history of particular 'familial and other emotional and desiring engagements' (Lennon, 2004, p. 117). As such, the way in which she makes sense of UTx-IVF demonstrates how specific experiences of those targeted by the innovation discern how social imaginaries of female embodiment entangle with UTx-IVF.

In the past few decades, women's engagements with medical interventions such as cosmetic surgery and medical AHR have been subject to discussions about whether these technologies obscure agency and consolidate normative understandings of female embodiment and motherhood (Davis, 2003; Lorentzen, 2008; Raymond, 1993). More recent work has considered medical practice to be co-constructed and situated. Demonstrating the material-discursive entanglement of the body and medical practices it has given rise to the argument that the materiality of physical processes and structures, embodied experiences, technological advances, and discursive understandings are inseparable from one another (Barad, 2003; Johnson, 2020; Murphy, 2012; Thompson, 2005). The research I have presented in this chapter feeds into such work, demonstrating the situatedness of the approach that interviewees take to their embodiment and to UTx-IVF as a technological innovation. The interviewees describe how shared social imaginaries of female embodiment—expressed, for example, in close relationships, public images, and writings—shape how they experience their bodies, as well as their expectations, desires, and imaginaries of their future (c.f. Hudson, 2020;

Lennon, 2004). The analysis highlights the importance of considering the situatedness of medical innovations, and individuals' approaches to them—in terms of both shared social imaginaries, and the individual histories of those to whom they apply.

Conclusion

UTx-IVF has developed rapidly over the past 20 years, with the ultimate goal of delivering healthy children. In this chapter, I aimed to examine the imagined futures that arise with the prospect of UTx-IVF, and how they are entangled with the accounts of women who live without uteri. I was interested in the expectations they had that went beyond reproduction, and in what they imagined that UTx-IVF would enable them to do or become. In the above discussion, I showed that the desire for options and choice expressed by my interviewees ties into a shared social imaginary of the promises of medical AHR. The potentiality of UTx-IVF thus includes not only a route to gestational and genetic parenthood, but also to the possibility of a future reproductive trajectory that is 'open to choice' (Taussig et al., 2013, p. s4), and aligned with reproductive trajectories that others are imagined to share. UTx-IVF was positioned as a technology that provisions agency, in as much as it offers the possibility of envisioning a future with increased (reproductive) options, choice, and control. I also considered how the interviewees made sense of their embodied histories of living without a vagina, and demonstrated how their experiences, desires, and approaches to UTx-IVF were entangled with social imaginaries of female embodiment. This analysis makes visible how the materiality of embodiment is shaped by conspicuous shared imaginaries and, as such, inhibits certain forms of agency.

Gender scholars, anthropologists, and medical sociologists have shown the importance of examining how medical AHR contributes to both reinforcement and questioning of norms that govern gendered embodiment. Such work has demonstrated the complex effects of medical AHR as both limiting and empowering—with authors often calling for a more nuanced picture of the efforts required by those involved in these practices, and for alternative social imaginaries of reproduction and female embodiment (Baker, 2004; Letherby, 2002; Peters et al., 2007; Sandelowski, 1990). However, as Lennon (2004, p. 120) asserts, drawing on work by Luce Irigaray, we cannot transform imaginaries simply by claiming that they are false: the alternative images must grab people's imaginations, 'take hold of people's lives and be found livable.' In terms of the wish expressed by the interviewees to have the same options as everyone else, and in light of their embodied histories, it is possible to construe UTx-IVF as stimulating an alternative and more livable imaginary of embodied and reproductive futures. At the same time, the accounts given by the women in this study highlight the effort involved in negotiating their reproductive trajectories and gendered embodiments. They describe ways in which medical AHR may make life more livable, without underestimating the difficulties involved. In presenting stories of medical AHR's impacts on the lives of individuals, in cases of both success and failure, the interviewees' accounts contribute to alternative imaginaries that provide a more nuanced understanding of what it might mean to wish for or undergo medical AHR. When seeking to create such alternative imaginaries, however, I suggest that it is necessary to avoid placing the responsibility for such transformations on the individuals who seek medical treatment for their infertility. Instead, we must make a collaborative effort to provide space in our shared social imaginaries for different kinds of stories about medical AHR. By appreciating that 'everyone else's' reproductive trajectories are not necessarily smooth, or simply a matter of individual preference and choice, such alternative imaginaries may provide people with more 'satisfying and livable ways of being in their environment' (Lennon, 2004, p. 120).

In closing, I look again at how my results feed into discussions of women's motivations for seeking UTx-IVF, emphasizing the urgency of considering how theoretical notions of women's motivations are deployed in ongoing debates around the procedure. In these debates, certain motivations are given priority over others, such as experiencing gestation (Lefkowitz et al., 2012, p. 444); personal or legal contraindications to surrogacy and adoption measures (Lefkowitz et al., 2012, p. 444); a 'desire to experience pregnancy'; and a 'desire to be a mother' (Bayefsky & Berkman, 2016, p. 355), all of which have been mobilized as criteria as for treatment access.

Since UTx-IVF is on the verge of being introduced into clinical practice, I suggest that we consider carefully how and why certain motivations are ascribed greater significance in ethical discussions and policy development, without disregarding infertility as such, the desire of many women to experience pregnancy, or their desire for genetic linkage to a future child as diagnostic categories. The accounts provided by the women in the study represent some of the many ways in which life without a uterus can be experienced, and the ways in which the prospect of acquiring a uterus can be imagined. As the technology travels and extends its grounds new challenges will arise and call, for example, for further examinations of the contextual and situated meanings attached to the uterus in relation to reproductive rights and citizenship across different jurisdictions. In light of this, the analysis in the study underlines the importance of considering the multitude of motivations and desires attached to UTx-IVF. This entails understanding women's motivations, not as fixed, but as entangled with their previous experiences and situated in shared social imaginaries of reproductive trajectories and female embodiment. In practice, such an understanding allows us to call into question previously assumed motivations as the basis for formulating policies and criteria for access and allocation (see also Allyse, 2018; Horsburgh, 2017; Richards et al., 2019) and to make care and support more sensitive to the specificities of affected women's embodied experiences and desires.

Notes

1. All names mentioned in the chapter are pseudonyms.
2. […] indicates that a sentence or section has been removed to ensure confidentiality or to increase readability (as in the case of stutters or sidetracks, for example).
3. This research has been funded by the Swedish Research Council (grant no. 2015-00972).
4. For other analyses of the material see Guntram (2021).
5. In Sweden, surrogacy is not specifically regulated. In practice, however, it is inaccessible since legislation stipulates that fertility treatment, such as

IVF, can only be accessed by single women who intend to parent the child and who can carry a pregnancy and give birth, and by couples who intend to parent the child and in which one of the individuals is a woman who can carry a pregnancy and give birth (Ministry of Health and Social Affairs, 2006). Consequently, IVF cannot be used in surrogate arrangements.

References

ACOG. (2002). ACOG committee opinion no. 247, July 2002: Nonsurgical diagnosis and management of vaginal agenesis. *International Journal of Gynecology & Obstetrics, 79*(2), 167–170.

Alghrani, A. (2018a). Uterus transplantation in and beyond cisgender women: Revisiting procreative liberty in light of emerging reproductive technologies. *Journal of Law and the Biosciences, 5*(2), 301–328. https://doi.org/10.1093/jlb/lsy012

Alghrani, A. (2018b). *Regulating assisted reproductive technologies: New horizons.* Cambridge University Press.

Allyse, M. (2018). "Whole again": Why are penile transplants less controversial than uterine? *The American Journal of Bioethics, 18*(7), 34–35. https://doi.org/10.1080/15265161.2018.1478044

Arora, K. S., & Blake, V. (2015). Uterus transplantation: The ethics of moving the womb. *Obstetrics & Gynecology, 125*(4), 971–974. https://doi.org/10.1097/AOG.0000000000000707

Baker, M. (2004). The elusive pregnancy: Choice & empowerment in medically assisted conception. *Women's Health and Urban Life: An International and Interdisciplinary Journal, 3*(1), 34–55. https://tspace.library.utoronto.ca/handle/1807/1056

Barad, K. (2003). Posthumanist performativity: Toward an understanding of how matter comes to matter. *Signs, 28*(3), 801–831. https://doi.org/10.1086/345321

Bayefsky, M. J., & Berkman, B. E. (2016). The ethics of allocating uterine transplants. *Cambridge Quarterly of Healthcare Ethics, 25*(3), 350–365. https://doi.org/10.1017/S0963180115000687

Becker, G. (2000). *The elusive embryo: How women and men approach new reproductive technologies.* University of California Press.

Beckman, L. J., & Harvey, S. M. (2005). Current reproductive technologies: Increased access and choice? *Journal of Social Issues, 61*(1), 1–20. https://doi.org/10.1111/j.0022-4537.2005.00391.x

Brännström, M. (2015). The Swedish uterus transplantation project: The story behind the Swedish uterus transplantation project. *Acta Obstetricia et Gynecologica Scandinavica, 94*(7), 675–679. https://doi.org/10.1111/aogs.12661

Brännström, M. (2019). Chapter 23: Uterine transplantation. In P. C. K. Leung & J. Qiao (Eds.), *Human reproductive and prenatal genetics* (pp. 515–525). Academic Press. https://doi.org/10.1016/B978-0-12-813570-9.00023-1

Brännström M., Johannesson L., Bokström H., et al. (2015). Livebirth after uterus transplantation. Lancet 385(9968): 607–616. https://doi.org/10.1016/S0140-6736(14)61728-1.

Brännström, M., Johannesson, L., Dahm-Kähler, P., Enskog, A., Mölne, J., Kvarnström, N., Diaz-Garcia, C., Hanafy, A., Lundmark, C., Marcickiewicz, J., Gäbel, M., Groth, K., Akouri, R., Eklind, S., Holgersson, J., Tzakis, A., & Olausson, M. (2014). First clinical uterus transplantation trial: A six-month report. *Fertility and Sterility, 101*(5), 1228–1236. https://doi.org/10.1016/j.fertnstert.2014.02.024

Brännström, M., Dahm Kähler, P., Greite, R., Mölne, J., Díaz-García, C., & Tullius, S. G. (2018). Uterus transplantation: A rapidly expanding field. *Transplantation, 102*(4), 569–577. https://doi.org/10.1097/TP.0000000000002035

Brännström, M., Boccio, M. V., & Pittman, J. (2020). Uterus transplantation: The science and clinical update. *Current Opinion in Physiology, 13*, 49–54. https://doi.org/10.1016/j.cophys.2019.10.004

Brännström, M., Enskog, A., Kvarnström, N., Ayoubi, J. M., & Dahm-Kähler, P. (2019). Global results of human uterus transplantation and strategies for pre-transplantation screening of donors. *Fertility and Sterility, 112*(1), 3–10. https://doi.org/10.1016/j.fertnstert.2019.05.030

Burns, M. (2003). I. Interviewing: Embodied communication. *Feminism & Psychology, 13*(2), 229–236. https://doi.org/10.1177/0959353503013002006

Catsanos, R., Rogers, W., & Lotz, M. (2013). The ethics of uterus transplantation. *Bioethics, 27*(2), 65–73. https://doi.org/10.1111/j.1467-8519.2011.01897.x

Davis, K. (2003). *Dubious equalities and embodied differences: Cultural studies on cosmetic surgery*. Rowman & Littlefield Publishers.

Dawney, L. (2011). Social imaginaries and therapeutic self-work: The ethics of the embodied imagination. *The Sociological Review,* 59(3), 535–552. https://doi.org/10.1111/j.1467-954X.2011.02015.x

Ellingson, L. L. (1998). 'Then you know how i feel': Empathy, identification, and reflexivity in fieldwork. *Qualitative Inquiry,* 4(4), 492–514. https://doi.org/10.1177/107780049800400405

EPN. (2014). Protocol, reference nr. 2014/436-31.

Exley, C., & Letherby, G. (2001). Managing a disrupted lifecourse: Issues of identity and emotion work. *Health,* 5(1), 112–132. https://doi.org/10.1177/136345930100500106

Farrell, R. M., Flyckt, R., & Falcone, T. (2018). The call for a closer examination of the ethical issues associated with uterine transplantation. *Journal of Minimally Invasive Gynecology,* 25(6), 933–935. https://doi.org/10.1016/j.jmig.2018.07.016

Farrell, R. M., Johannesson, L., Flyckt, R., Richards, E. G., Testa, G., Tzakis, A., & Falcone, T. (2020). Evolving ethical issues with advances in uterus transplantation. *American Journal of Obstetrics and Gynecology,* 222(6). https://doi.org/10.1016/j.ajog.2020.01.032

Franklin, S. (1997). *Embodied progress: A cultural account of assisted conception.* Routledge.

Franklin, S. (2013). *Biological relatives: IVF, stem cells and the future of kinship.* Duke University Press.

Franklin, S.B. and McKinnon, S. (2001). Relative Values: Reconfiguring Kinship Studies. Durham, NC: Duke University Press.

Gammeltoft, T. M., & Wahlberg, A. (2014). Selective reproductive technologies. *Annual Review of Anthropology,* 43(1), 201–216. https://doi.org/10.1146/annurev-anthro-102313-030424

Gatens, M. (1996). *Imaginary bodies: Ethics, power, and corporeality.* Routledge.

Gatens, M., & Lloyd, G. (1999). *Collective imaginings: Spinoza, past and present.* Routledge.

Gunnarson, M. (2016). *Please be patient: A cultural phenomenological study of haemodialysis and kidney transplantation care.* PhD thesis, Lund Studies in Art and Cultural Sciences. http://lup.lub.lu.se/record/8410361

Guntram, L. (2013). "Differently normal" and "normally different": Negotiations of female embodiment in women's accounts of 'atypical' sex development. *Social Science & Medicine,* 98, 232–238. https://doi.org/10.1016/j.socscimed.2013.09.018

Guntram, L. (2018). Hooked on a feeling? Exploring desires and 'solutions' in infertility accounts given by women with 'atypical' sex development. *Health, 22*(3), 259–276. https://doi.org/10.1177/1363459317693403

Guntram, L. (2021). May I have your uterus? The contribution of considering complexities preceding live uterus transplantation. *Medical Humanities*, 1–13. https://pubmed.ncbi.nlm.nih.gov/33627444/. Epub 2021 February 24.

Herlin, M., Bjørn, A.-M. B., Rasmussen, M., Trolle, B., & Petersen, M. B. (2016). Prevalence and patient characteristics of Mayer–Rokitansky–Küster–Hauser syndrome: A nationwide registry-based study. *Human Reproduction, 31*(10), 2384–2390. https://doi.org/10.1093/humrep/dew220

Holt, R. E., & Slade, P. (2003). Living with an incomplete vagina and womb: An interpretative phenomenological analysis of the experience of vaginal agenesis. *Psychology, Health & Medicine, 8*(1), 19. https://doi.org/10.1080/1354850021000059232

Horsburgh, C. C. (2017). A call for empirical research on uterine transplantation and reproductive autonomy. *Hastings Center Report, 47*(S3), S46–S49. https://doi.org/10.1002/hast.795

Hudson, N. (2020). Egg donation imaginaries: Embodiment, ethics and future family formation. *Sociology, 54*(2), 346–362. https://doi.org/10.1177/0038038519868625

Järvholm, S., Johannesson, L., & Brännström, M. (2015). Psychological aspects in pre-transplantation assessments of patients prior to entering the first uterus transplantation trial. *Acta Obstetricia et Gynecologica Scandinavica, 94*(10), 1035–1038. https://doi.org/10.1111/aogs.12696

Johannesson, L., Kvarnström, N., Mölne, J., Dahm-Kähler, P., Enskog, A., Diaz-Garcia, C., Olausson, M., & Brännström, M. (2015). Uterus transplantation trial: 1-year outcome. *Fertility and Sterility, 103*(1), 199–204. https://doi.org/10.1016/j.fertnstert.2014.09.024

Johnson, E. (2020). *Refracting through technologies: Bodies, medical technologies and norms*. Routledge.

Jones, B. P., Williams, N. J., Saso, S., Thum, M.-Y., Quiroga, I., Yazbek, J., Wilkinson, S., Ghaem-Maghami, S., Thomas, P., & Smith, J. R. (2019). Uterine transplantation in transgender women. *BJOG: An International Journal of Obstetrics & Gynaecology, 126*(2), 152–156. https://doi.org/10.1111/1471-0528.15438

Kroløkke, C., & Kotsi, F. (2019). Pink and blue: Assemblages of family balancing and the making of Dubai as a fertility destination. *Science, Technology, & Human Values, 44*(1), 97–117. https://doi.org/10.1177/0162243918783059

Kvarnström, N., Järvholm, S., Johannesson, L., Dahm-Kähler, P., Olausson, M., & Brännström, M. (2017). Live donors of the initial observational study of uterus transplantation: Psychological and medical follow-up until 1 year after surgery in the 9 cases. *Transplantation, 101*(3), 664–670. https://doi.org/10.1097/TP.0000000000001567

Lefkowitz, A., Edwards, M., & Balayla, J. (2012). The Montreal criteria for the ethical feasibility of uterine transplantation. *Transplant International, 25*(4), 439–447. https://doi.org/10.1111/j.1432-2277.2012.01438.x

Lennon, K. (2004). Imaginary bodies and worlds. *Inquiry, 47*(2), 107–122. https://doi.org/10.1080/00201740410005132

Letherby, G. (2002). Challenging dominant discourses: Identity and change and the experience of 'infertility' and 'involuntary childlessness'. *Journal of Gender Studies, 11*(3), 277–288. https://doi.org/10.1080/0958923022000021241

Lie, M., & Lykke, N. (2016). *Assisted reproduction across borders: Feminist perspectives on normalizations, disruptions and transmissions*. Routledge.

Lorentzen, J. M. (2008). 'I know my own body': Power and resistance in women's experiences of medical interactions. *Body & Society, 14*(3), 49–79. https://doi.org/10.1177/1357034X08093572

Mabel, H., Farrell, R. M., & Tzakis, A. G. (2018). On gender and reproductive decision-making in uterine transplantation. *The American Journal of Bioethics, 18*(7), 3–5. https://doi.org/10.1080/15265161.2018.1489655

Mamo, L. (2007). *Queering Reproduction: Achieving Pregnancy in the Age of Technoscience*. Durham, N.C.: Duke University Press.

Mertes, H., & Assche, K. V. (2018). UTx with deceased donors also places risks and burdens on third parties. *The American Journal of Bioethics, 18*(7), 22–24. https://doi.org/10.1080/15265161.2018.1478029

Ministry of Health and Social Affairs (2006). Lag (2006: 351) Om Genetisk Integritet m.m. Swedish Code of Statutes.

Murphy, M. (2012). *Seizing the means of reproduction: Entanglements of feminism, health, and technoscience*. Duke University Press.

Nordqvist, P. (2008). Feminist heterosexual imaginaries of reproduction. *Feminist Theory 9*(3) 273–292. https://doi.org/10.1177/1464700108095851

Papachristou, C., Walter, M., Schmid, G., Frommer, J., & Klapp, B. F. (2009). Living donor liver transplantation and its effect on the donor-recipient relationship: A qualitative interview study with donors. *Clinical Transplantation, 23*(3), 382–391.

Peters, K., Jackson, D., & Rudge, T. (2007). Failures of reproduction: Problematizing 'success' in assisted reproductive technology. *Nursing Inquiry, 14*(2), 125–131. https://doi.org/10.1111/j.1440-1800.2007.00363.x

Raymond, J. G. (1993). *Women as wombs: Reproductive technologies and the battle over women's freedom.* Spinifex Press.

Richards, E. G., Agatisa, P. K., Davis, A. C., Flyckt, R., Mabel, H., Falcone, T., Tzakis, A., & Farrell, R. M. (2019). Framing the diagnosis and treatment of absolute uterine factor infertility: Insights from in-depth interviews with uterus transplant trial participants. *AJOB Empirical Bioethics, 10*(1), 23–35. https://doi.org/10.1080/23294515.2019.1572672

Riggs, D. W., & Due, C. (2013). Representations of reproductive citizenship and vulnerability in media reports of offshore surrogacy. *Citizenship Studies, 17*(8), 956–969.

Sandelowski, M. (1990). Fault lines: Infertility and imperiled sisterhood. *Feminist Studies, 16*(1), 33–51. https://doi.org/10.2307/3177955

Sandelowski, M., Harris, B. G., & Holditch-Davis, D. (1991). "The clock has been ticking, the calendar pages turning, and we are still waiting": Infertile couples' encounter with time in the adoption waiting period. *Qualitative Sociology, 14*(2), 147–173. https://doi.org/10.1007/BF00992192

Sandman, L. (2018). The importance of being pregnant: On the healthcare need for uterus transplantation. *Bioethics, 32*(8), 519–526. https://doi.org/10.1111/bioe.12525

Saso, S., Clarke, A., Bracewell-Milnes, T., Saso, A., Al-Memar, M., Thum, M.-Y., Yazbek, J., Priore, G. D., Hardiman, P., Ghaem-Maghami, S., & Smith, J. R. (2016). Psychological issues associated with absolute uterine factor infertility and attitudes of patients toward uterine transplantation. *Progress in Transplantation, 26*(1), 28–39. https://doi.org/10.1177/1526924816634840

Scheper-Hughes, N. (2007). The tyranny of the gift: Sacrificial violence in living donor transplants. *American Journal of Transplantation, 7*(3), 507–511. https://doi.org/10.1111/j.1600-6143.2006.01679.x

Taussig, K.-S., Hoeyer, K., & Helmreich, S. (2013). The anthropology of potentiality in biomedicine: An introduction to supplement 7. *Current Anthropology, 54*(S7), 3–14. https://doi.org/10.1086/671401

Testa, G., Koon, E. C., & Johannesson, L. (2017a). Living donor uterus transplant and surrogacy: Ethical analysis according to the principle of equipoise. *American Journal of Transplantation, 17*(4), 912–916. https://doi.org/10.1111/ajt.14086

Testa, G., Koon, E. C., Johannesson, L., McKenna, G. J., Anthony, T., Klintmalm, G. B., Gunby, R. T., Warren, A. M., Putman, J. M., dePrisco, G., Mitchell, J. M., Wallis, K., & Olausson, M. (2017b). Living donor uterus transplantation: A single center's observations and lessons learned from early setbacks to technical success. *American Journal of Transplantation, 17*(11), 2901–2910. https://doi.org/10.1111/ajt.14326

Thompson, C. (2005). *Making parents: The ontological choreography of reproductive technologies.* MIT Press.

Throsby, K. (2004). Negotiating normality when IVF fails. In M. Bamberg & M. Andrews (Eds.), *Considering counter narratives: Narrating, resisting, making sense* (pp. 61–82). John Benjamins.

Ulrich, M., & Weatherall, A. (2000). Motherhood and infertility: Viewing motherhood through the lens of infertility. *Feminism & Psychology, 10*(3), 323–336. https://doi.org/10.1177/0959353500010003003

Whittaker, A. M. (2011). Reproduction opportunists in the new global sex trade: PGD and non-medical sex selection. *Reproductive BioMedicine Online, 23*(5), 609–617. https://doi.org/10.1016/j.rbmo.2011.06.017

Wilkinson, S., & Williams, N. J. (2016). Should uterus transplants be publicly funded? *Journal of Medical Ethics, 42*(9), 559–565. https://doi.org/10.1136/medethics-2015-102999

Zeiler, K., & Guntram, L. (2014). Sexed embodiment in atypical pubertal development: intersubjectivity, excorporation, and the importance of making space for difference. In L. F. Käll & K. Zeiler K. (Eds.), *Feminist phenomenology and medicine* (pp. 141–160). SUNY Press.

7

Sized Out: Fatness, Fertility Care, and Reproductive Justice in Aotearoa New Zealand

George Parker and Jade Le Grice

Introduction

Reproductive justice is a framework for thought and action that links human reproductive rights with broader issues of social justice (Ross & Solinger, 2017). First articulated by Black, Indigenous and people(s) of colour (BIPOC), reproductive justice challenges the tendency of white Western feminists to focus on the right to end unwanted pregnancies to the exclusion of other concerns surrounding reproductive rights. Reproductive justice emphasises the relationship between reproductive decision-making and the social, cultural, and political structures that determine those decisions, highlighting the limitations of the very notion

G. Parker (✉)
Te Herenga Waka—Victoria University of Wellington,
Wellington, New Zealand
e-mail: george.parker@vuw.ac.nz

J. Le Grice
University of Auckland, Auckland, New Zealand

of reproductive *choice*. By attending to the broader conditions of people's lives, reproductive justice advocates claim that all people with reproductive capacity require a safe and dignified environment which upholds their right to have a child, their right *not* to have a child, and their right to parent children in safe and healthy environments (Ross & Solinger, 2017, p. 9).

This expanded vision for reproductive politics provides a much-needed framework for considering equity issues in relation to assisted reproductive technologies, by raising questions about who can access these technologies in order to fulfil the right to have a child. In this chapter, we draw on the framework of reproductive justice to critically assess the implications of Aotearoa New Zealand's Clinical Priority Access Criteria (CPAC) that determine access to publicly funded fertility treatment based on a range of specifications, including the Body Mass Index (BMI) of the person seeking treatment. CPAC criteria were introduced in Aotearoa New Zealand with the goal of prioritising publicly funded fertility treatment for those who are most likely to benefit from it (Farquhar & Gillett, 2006). Weight-based restrictions are based on medical assumptions about fatness as a modifiable lifestyle factor that reduces fertility and contributes to complications in pregnancy and beyond (Farquhar & Gillett, 2006). The assumption is that fat people experiencing infertility will be incentivised to lose weight, which may then resolve fertility issues, increase the effectiveness of fertility treatment, and reduce the risk of complications in the pregnancies that result from fertility treatment (e.g., Pandey et al., 2010).

However, the assumptions that underpin weight-based criteria for access to publicly funded fertility treatments are deeply contested within and beyond the medical community. Furthermore, existing medical debate about fat[1] women's[2] access to assisted fertility treatment pays little attention to the perspectives of fat women themselves. This chapter presents a critical discussion about the implications of including BMI in CPAC for fertility treatment. Our discussion is informed by an affective-discursive analysis of qualitative interviews with six ethnically diverse, self-identified fat women who were unable to access fertility care because of their size. Key themes drawn from participants' accounts demonstrate how the stigmatising and demoralising effects of weight-based exclusion from fertility treatment compound the distress caused by infertility. Participants engaged in self-blame and took responsibility for their predicament in ways that were disruptive to their health and wellbeing. However, participants also resisted and challenged medical assumptions about fatness and fertility. Weight-based criteria in fertility services resulted in reproductive gatekeeping on the basis of race and class, which is actively harmful to fat people seeking fertility treatment and their families. Here, the principles of reproductive justice provide a counterpoint by which to reimagine equitable access to fertility care.

Rationing Assisted Reproduction in Aotearoa: Clinical Priority Access Criteria and BMI

The CPAC for infertility was implemented in 2000, as part of a broader project to ration elective, publicly funded procedures by allowing patients to be ranked for preferential treatment (Farquhar & Gillett, 2006). As a result of the introduction of the CPAC for publicly funded fertility treatment, women would only be eligible for treatment if their Body Mass Index (BMI) is inside the range of 18–32 kg/m^2. Women with a BMI higher than 32 were given a stand-down period, with treatment only commencing if a lower BMI was achieved through weight loss (Gillett & Peek, 1997). For many women with a BMI higher than 32 kg/m^2, this has prevented them from accessing public fertility services, with general practitioners (GP)[3] declining to refer, or fertility services themselves declining referrals. Since 2018, some adjustments to the criteria have permitted women to be seen for assessment by fertility specialists if their BMI is less than 35, but unless their BMI is less than or equal to 32, medical practitioners continue to decline publicly funded treatment (National Women's Health, 2020). Aotearoa New Zealand was one of the first countries in the world to introduce weight eligibility criteria for publicly funded fertility treatment, although a number of other countries have since followed suit, including the United Kingdom and Canada with some controversy (Friedman, 2015).

This exclusionary access to publicly funded fertility treatment is stratified along ethnic and socio-economic lines. The New Zealand Health Survey 2018/2019 describes around a third of adults in Aotearoa New Zealand as 'obese,' with higher prevalence among Pacific people (66.5%) and Māori (48.3%) compared to NZ European (29.1%) (Ministry of Health, 2020). Further, adults living in the most socioeconomically deprived areas were significantly more likely to be classified 'obese' than adults living in the least deprived areas (Ministry of Health, 2019). Despite this, the same CPAC model and the BMI threshold were used for all ethnic groups, without recognition of these socio-economic disparities (Gillett et al., 2006). Gillett et al. (2006, p. 1221) observe that this may seem "harsh" for Māori and Pacific[4] women; however, such implications were considered to be outweighed by the public health benefits of incentivising weight reduction—in particular, reducing the burden of obesity on healthcare resources (Farquhar & Gillett, 2006).

When considering the momentum behind medical concern with obesity as a health issue in Global North countries over the past two decades, the significance given to fatness in fertility treatment is hardly surprising. Growing medical concern with rates of fatness in the population, its effects on health, and the projected associated health care costs have precipitated the emergence of a weight-based paradigm in healthcare (O'Hara & Gregg, 2006). This in turn has led to what O'Hara and Gregg call a "war on obesity" in social, educational, and healthcare institutions. The key assumptions that underpin this weight-based paradigm include the centrality of fatness to the health status of individuals, the conception of fatness as a modifiable lifestyle factor that results primarily from energy imbalance in the individual, and the belief that weight loss is a realistic and achievable goal which will result in improved health (O'Hara & Gregg, 2006). These assumptions have informed wholesale government and health system efforts to encourage weight loss in the population (e.g. Gard & Wright, 2005).

The effects of fatness on reproductive health have been of particular concern to the medical community in recent years (Parker, 2014). The reported health risks (and healthcare costs) of so-called maternal obesity on both fertility and childbirth outcomes have occupied news headlines and led to a reallocation of public investment in weight-loss programmes from the general population to those targeted at pregnant people and new mothers (Parker, 2014). This has been fuelled by a swathe of medical science studies associating fatness with infertility, increased risk of perinatal complications, and long-term health effects on babies born to fat people (e.g. Poston et al., 2016). Indeed, fatness has been associated with an increased risk of almost all perinatal complications, including miscarriage, stillbirth, the occurrence of congenital abnormalities, hypertension, gestational diabetes, caesarean section, postpartum haemorrhage, and admission to the neonatal unit. In addition, the offspring of fat people have been argued to be at greater risk of a range of health complications, from adult fatness to asthma and autism (e.g., Armitage et al., 2008).

Public dialogue about maternal fatness frequently emphasises the costs fat people pose to reproductive and maternal health services. Consistent with the weight-based paradigm in healthcare, it proffers solutions grounded in the notion of individual responsibility. The weight-based paradigm in healthcare has been embedded across reproductive health services from access to fertility treatment, the medicalised management of fat pregnancies including restricted choice about birth plan and place, and weight-management programmes aimed at fat pregnant people and new mothers (Parker, 2014). So extensive is the problematisation of fatness in relation to reproduction that interventions such as the CPAC for fertility treatment are justified as a moral good regardless of any harms that may be attributed to them. As Farquhar and Gillett (2006, p. 1108) argue: "By encouraging lifestyle changes such as weight loss, the message that obesity is a major health problem is reinforced."

Weighing in: Critical Perspectives on Weight Restrictions for Fertility Treatment

While the weight-based paradigm in healthcare and the "war on obesity" have dominated the health landscape for the past two decades, there are many detractors. Critical obesity scholars across a range of disciplines have questioned the core assumptions that conflate fatness and ill-health, asking whether sustainable weight loss is realistic and even desirable for many fat people, and point to ways in which the legitimisation and amplification of fat hatred actually harms fat people's health (e.g., Bacon & Aphramor, 2011). Further, critical scholars have observed the utility of obesity in Global North countries to justify neoliberal government agendas aimed at reducing responsibility for, and involvement in, the conditions that determine the wellbeing of their people by responsibilising individuals for their own self-management (e.g., LeBesco, 2011). Accepted as a major determinant of health that is directly under individual control, some have argued that obesity is an ideal metric for displacing the responsibility for health and wellbeing from the state to individual citizens (LeBesco, 2011). Positioned as 'to blame' for their predicament, fat people can be subject to sanction without consideration as to how social factors, forms of marginalisation, and limited access to resources actively shape the body weight and health of individuals. This is particularly salient for fat people of colour and Indigenous people who live in poverty and are subject to the compounded effects of racism, colonisation, and socio-economic marginalisation (LeBesco, 2011).

More recently, critical literature has focused on the specific harms of fat shame and blame deployed in reproductive health services, where fat people become framed not only as unworthy citizens but as unworthy parents (or parents-to-be) (e.g., Bombak et al., 2016; Friedman, 2015; LaMarre et al., 2020; Parker & Pausé, 2019; Ward & McPhail, 2019). Scholars have pointed to the ways in which medical concern with fatness in reproduction intersects with the dynamics of mother-blame and foetal protectionism in the West (Herdon, 2018). This works to position fat people who are, or want to be, pregnant as both an unacceptable 'drain' on the health system *and* as 'undeserving' of their pregnancies by intentionally risking harm to their babies. Fat people experiencing infertility, who are currently pregnant, or who are new parents, have found that this framing can rob them of the joy of pregnancy (Parker & Pausé, 2019) or compound the distress of infertility (LaMarre et al., 2020), causing harm to their own health, and/or that of their potential or existing child/ren.

The problematisation of fat reproductive bodies is gendered, attributing responsibility for reproductive health to women and gender-diverse people who can carry pregnancies and ignore the structural inequalities that determine reproductive health outcomes. However, critical scholars working through the lens of reproductive justice have also pointed to the ways in which existing raced and classed biases about who is (and isn't) considered 'fit' to reproduce and parent are amplified (e.g., Sanders, 2019). Contemporary obesity-related policies and practices that disproportionately affect the reproductive autonomy and dignity of Indigenous women and women of colour reproduce racist, colonial, and eugenicist ideas that curtail the reproductive potential of Indigenous women and women of colour for 'the greater good' (Parker et al., 2019). These ideas have been, and continue to be, embedded in policies and practices that undermine the reproductive dignity and autonomy of Indigenous women and women of colour (e.g., Ware et al., 2017). Raced, classed, and gendered marginalisation are conveniently masked by the individualised discourse of self-management, which suggests that nobody is to blame for the issues associated with fatness other than fat people themselves (Parker et al., 2019).

The Study

In this chapter, we draw on the findings of qualitative interviews undertaken with six self-identified fat, ethnically diverse, cisgender women located in Auckland,[5] Aotearoa New Zealand, who were excluded from fertility treatment on the basis of their BMI. The interviews were collected as part of a larger study on maternal obesity that included interviews with 27 self-identified fat women about their fertility, pregnancy, and childbirth experiences. While the six interviews constitute a small sample, Braun and Clarke (2013) note the acceptability, indeed the desirability, of a small sample size in experience-focused qualitative research. This allows the researcher to retain a focus on individuals' experiences whilst also convincingly demonstrating patterns across the data set. We note that the six interviews demonstrated remarkable consistency in the thematic patterns identified in analysis.

The research is informed by Bacchi's (2012, p. 1) theoretical concept of problematisation. Problematisation is a concept that draws on Foucauldian-inspired poststructuralist theory putting into question accepted truths. The goal of research informed by this approach is to examine how and why certain things (e.g., behaviour and certain kinds of bodies) become a problem, how these problems are taken up in the process of subjectification as "truth," to what effect, and whose political interests are served. Applied to this research focus, problematisation seeks to trouble dominant medical assumptions about fat reproductive bodies, and to illuminate the practices, political structures, and other strategic relations that constitute such knowledges and give them their power (Bacchi, 2012). We examine how medical knowledges about reproduction, reified as 'truths,' shape and constrain the conditions of possibility for fat reproductive subjects. The study of problematisations is not limited to critique, and also seeks to highlight non-inevitability by generating more diverse and equitable ways of knowing and being, beyond those offered by dominant discourse (Gavey, 2011, p. 185).

Participants were recruited through articles in local newspapers and social media networks. Of the six participants whose interviews are included in this chapter, two participants were Māori, two were Pacific, and two were New Zealand European. Four participants described the experience of primary infertility prior to the birth of a child, and two described the experience of secondary infertility, with both already having given birth to one child each. In order to de-centre medical categorisations of fatness such as BMI, and to ensure that negative experiences with evaluations of weight in healthcare encounters were not replicated in the study, self-identification of fatness was a core principle of study recruitment. All six participants were experiencing (or had experienced) infertility and had been unable to secure a referral from their GP or specialist for publicly funded fertility assessment and treatment as a result of the CPAC criteria related to BMI. Ethical approval for this study was gained from the University of Auckland Human Participants Ethics Committee (91/68).

Interviews were recorded, pseudonyms were given, and prepared transcripts were analysed using affective-discursive analysis that incorporated the principles of intersectionality. Affective-discursive practice is an analytical approach that combines critical discourse analysis with attention to affect, the rich emotional dimensions of participant interviews. The critical discourse analyst is interested in identifying discourses (i.e., regulated systems of statements); the ways in which they constitute knowledges or truths; and the ways in which these knowledges or truths regulate practices, experiences, and subjectivities (Braun & Clarke, 2013). Taking up the challenge of affect theory, affective-discursive practice attends to the enmeshment of discourse and affect (or feelings/emotions) in accounts produced by the meaning-making subject (Wetherell et al., 2015). The affective-discursive analyst is interested in the ways in which dominant discourses achieve their subjectifying effects through the energisation of affectivity; that is, by making the subject feel things such as disgust, fear, and shame (Bjerg & Staunæs, 2011). We refer to these effects as 'affective impacts' in the findings presented below.

Additionally, the principles of intersectionality[6] were incorporated into the analysis (Price, 2011). For the purposes of this discussion, intersectionality means attending to the ways in which the effects of discourse and the processes of subjectification are differentiated by participants' location within various axes of social power and by the effects of colonisation, racism, heterosexism and ciscentrism, economic marginalisation, ableism, sizeism, and ageism. The results of analysis are presented below grouped into three overarching themes. The themes were derived through a process of intensive theoretically informed reading of the transcripts, manual coding of the transcripts, and the collation of codes into candidate themes and sub-themes, which, through a recursive process, resulted in the capturing of three overarching themes that cohere together (Braun & Clarke, 2013).

"It's just all the heartbreaks and you're on your own": Gatekeeping Fertility Treatment

The interviews in this study highlighted a heavy affective burden that can result with being denied fertility treatment due to weight. The experience of infertility alone has been described as emotionally, psychologically, and spiritually taxing, incurring a sense of worthlessness; inadequacy and self-blame; loss of control; anger and resentment; grief and depression; lower life satisfaction; and a sense of isolation (e.g., Greil et al., 2010). Indeed, infertility has been described as an "emotional roller coaster" (Greil et al.,

2010, p. 144). For participants in this study, weight-based exclusion compounded and heightened the negative affectivity of infertility in particularly harmful ways.

Participants shared versions of a similar routine, periodically visiting their GP with concerns about their infertility, being weighed, and, when they did not meet the BMI threshold for referral to fertility services, being told to go away and lose weight. As Mere described:

> It's been five years now, and I have been to the doctors so many times asking them why I can't get pregnant. I've got a child, and I don't consider myself that big and it's just disheartening that they keep telling me to lose weight, they just say that is the reason why, go away and lose weight.

Participants described how their weight seemed to obscure their doctors' ability to consider other possible causes of their infertility, with little or no inquiry into participants' health or suggestions for other investigative options. The influence of anti-fat bias on the treatment provided by health care providers to fat people has been comprehensively demonstrated in research (e.g., Chrisler & Barney, 2017). Participants described how their doctors refused to consider alternative possibilities regardless of whether the participant's weight remained consistent, or whether it was on the cusp of the BMI threshold for referral. As Mere described, "not once did they look at my history and notice that my weight is quite steady. They just look at the number, and say if you want it enough, you'll go away and lose weight." Similarly, Natia said:

> It's just you need to lose weight. And I'm like, something down there isn't working, and nobody wants to investigate it unless I'm a certain weight. But yeah, I believe that whole focus on my weight is ignoring something internal that is going on. It's not fair.

Only two of the six participants in this study had ever secured a referral from their GP to fertility services, despite experiencing infertility over several years, and both were declined to be seen by fertility services because of their BMI. Alison described receiving a letter back from fertility services saying that her case had been reviewed, resulting in the recommendation of weight-loss surgery before she would be eligible for treatment. Alison described this experience:

> There was this tone to [the letter] saying we wouldn't even consider you at this weight, you know I think it actually said, we would advise you don't get pregnant at this weight, something like that. It didn't feel good at all, it felt like, it almost felt like there was a grand committee of people sitting around deciding who was allowed to get pregnant and who is not, like gatekeeping—you know?

This perception that the BMI criteria functioned as a form of gatekeeping was common across the accounts. After four years of infertility, Natia finally convinced her GP to send a referral to fertility services, but, like Alison, was also declined an appointment. Natia explained:

> I got the letter saying because of your weight, you know, we can't proceed any further, when you get to a healthy weight range then we'll see you, all because of six kilos. And it feels like something just out of reach, I'm asking for help and not receiving it.

Consistent with the weight-based paradigm in health, GPs and/or fertility services emphasised participants' individual responsibility for their weight, foregrounding the associated weight loss project. Participants described their doctors' emphasis on weight loss as the "solution" to their infertility, with the assumption that their fatness could simply be corrected through self-discipline and the willpower to make the right lifestyle choices about diet and exercise. As Natia described, the deferral of responsibility from fertility care to the participants themselves exacerbated the self-blame associated with infertility:

> It's kind of like all the messages about losing weight, it makes it really up to you. You know, you hold all the cards, so if you don't do anything about it then, you know, it's kind of your own doing. It's kind of like it's heavy, you know, it's kind of like a heavy burden. And yeah, it's willpower, you know, have I got the willpower?"

Participants described the effect of this reproductive healthcare treatment as being akin to abandonment by health services in relation to their infertility. Participants' self-blame for their experience of infertility was compounded by the lack of information, support, and resources to assist them with weight loss, or even an explanation as to why it was considered necessary. Mere described this disjuncture in reproductive healthcare:

> All they care about is telling you to lose weight. If they really cared, you know, they could have actually given us more advice, recommendations, you know, just something that would help you out rather than shutting you down by saying lose the weight, because it's like straight away you switch off and just feel down.

Natia's reflections were very similar to Mere's:

> The information they give you, it's been really vague. They think that weight loss is the answer to my whole problem when maybe it's not. You know you go to health services because you're in trouble, or you've got complications, but you're not getting any help, it's kind of heart-breaking.

Lacking information about how to tackle weight loss, coupled with self-blame for their predicament, participants described a sense of hopelessness and despair about how to find a pathway out of their infertility. As Mere reflected on her experience of resignation:

> So, I just gave up asking because, I mean, they know what's out there and what's available, but I guess it comes down to my weight and that doesn't allow them to offer that help to me. It was like, why should they give that treatment to me when I'm big?

"There's a lot of judgement in our system": Reproducing Inequity, Stigma and Shame

Exclusion from publicly funded fertility services led participants to question how decisions were made about who was, and was not, worthy of having a baby, through the investment of public health dollars. While participants blamed themselves for their infertility, they were also critical of BMI exclusionary criteria as a fat shaming and racialised instrument. Fat shame and discrimination is widespread in the Global North context and has oppressive effects on the psychological, spiritual, and physical health of fat people (Gard & Wright, 2005). Fat shaming in healthcare has been shown to be particularly deleterious to the health and wellbeing of fat people (Ward & McPhail, 2019). Women experiencing infertility are already primed for self-blame; participants described how the problematisation of their fatness exacerbated this response, leading them to question whether or not they were worthy of having a baby. As Isabelle described:

> With regards to fertility, it's a super sensitive subject. You're dealing with miscarriage, pregnancy loss, the heart break over never falling pregnant, you know, every woman's birth right to have a baby isn't there for everybody, so you're dealing with some pretty intense emotional issues, so then on top of that to say "not only are you barren darling, but also you're fat and disgusting, ok?"

The destabilising experience of fat shaming in healthcare encounters, coupled with infertility, and the abandonment associated with fertility 'care' in response to weight-loss recommendations, led participants to resist and reject the perceived wisdom and expertise of their GP. Some participants raised questions about fat stigma, and the eugenic implications of denying access to fertility services on the basis of weight. Talia described distrusting her doctors' focus on her weight and, while investigating potential causes of her infertility, reflected:

You know, I didn't know if it was the doctors just not wanting me to conceive or something because you know I just felt like I'd been put aside, just an overweight person who shouldn't have children.

Alison likened her experience of receiving the letter declining her referral to fertility services like an episode of the TV show Grey's Anatomy, where they have "50 people in a room discussing your case." She concluded, "It felt like there was a bit of a class system as to who was allowed to get pregnant in New Zealand, and who was welcome."

Māori and Pacific participants, considering their exclusion through a lens of racial inequality, wondered whether the gatekeeping of fertility treatment was ultimately more about constraining brown women's ability to have babies. The colonial legacy of constraining the reproductive potential of brown women continues in contemporary New Zealand, manifesting in social policy and media vilification of young Māori and Pacific sole parents, young parents, and recipients of welfare (e.g., Ware et al., 2017). Several participants observed that BMI criteria for fertility treatment were policed more strictly in South Auckland, with its large Māori and Pacific population, compared to areas of the city with predominantly European populations. Participants described knowing white women who were referred to fertility services despite being over the BMI threshold. Talia, for example, wrestled with her sense that the BMI criteria was racialised, while simultaneously taking up the discourse of responsibility and self-blame:

I do wonder if they are just finding ways to save money. I think they are. And you know, it's hard to say, but I actually think its possibly an ethnicity thing… so we don't use their resources. And that's just not fair, yeah, really not fair. It's terrible. But yeah but I try not to look at it from that point of view, I guess because I do understand that it does come down to a choice, your choices, what choices you make in life in terms of your lifestyle, what you eat, how you live your life, everything.

Unpacking the racialised implications of BMI criteria for access to publicly funded fertility treatment, participants pointed to the ways in which BMI as a system of measurement disadvantages Māori and Pacific people with their higher mean body size. As Mere argued:

> They don't take account of different body frames. Being a Pacific Islander, it's a different body frame. They just assume we are all the same and make the BMI number for everybody. And I mean, they know all the facts about percentages of people in this area, and that we don't fit their BMI requirements, so yeah, maybe they're looking to save money by stopping brown people who are trying to get pregnant.

A Eurocentric valorisation of slender embodiment, and its conflation with health, disregards Māori cultural meanings and embodied understandings of health and reproductive potential. This creates a disjuncture between medicalised contexts and whānau contexts which are oriented to shared whakapapa[7] and normalisation of large-sized reproductive bodies across generations (Parker et al., 2019). As Natia reflected:

> When I'm in my family environment they're like there's nothing wrong with you, and then I come to the professionals and they're like you're overweight and that's why it's [pregnancy] not happening, and it's like I have to change hats.

Dominant medicalised discourses that problematise fatness in relation to infertility are argued to be individualising, fat shaming, and racist, all of which perpetuate the injustices of colonialism (Friedman, 2015; LaMarre et al., 2020; Parker et al., 2019). Through an intersectional lens Māori and Pacific women must navigate *both* the effects of fat shaming *and* the ways these meanings perpetuate and amplify racialised and eugenic ideas that they are less entitled to reproduce than European women. Furthermore, in this study, these discourses bring to light the stark failure of current public health policy to motivate weight reduction amongst those seeking fertility services. Instead, participants in our study described the deleterious effects of these discourses on their psychological, spiritual, and physical health. All the participants in this study described how weight messaging grafted to fat-shaming discourses, and self-blame associated with infertility, leading to feelings of guilt, shame, and self-hatred. As Mere described,

> They always just tell you that you're obese. And I'm like really? I just don't see myself as that. So, one of the doctors showed me a chart and where I was on it and I just thought, "oh, I hate myself quite a bit." I just hate myself. I felt down and then like too embarrassed to tell my partner that the problem is me. That I'm the one that's got issues, that I'm the reason we can't have any more kids.

Being made to feel bad about their weight was not health promoting for the participants in our study. Participants indicated that the affective impacts of these weight-exclusionary infertility healthcare encounters, experienced within a broader context of racialisation and fat shaming, led to a range of harmful effects including loss of connection with others, stress, depression, anxiety, self-medicating with alcohol and binge eating, and a diminished sense of agency over their health. Talia, for example, described:

> I think that is how my depression kicked in. I said "I give up, if no one is going to listen to me, the doctor isn't, the specialist isn't, what is the point, you know. Obviously, I'm not going to have children, no one's taking me seriously, no one's going to forward me to a fertility clinic. And then my binge eating came in because I was so depressed and I gave up, so that is when I started self-medicating.

Mere described a reversal in her personal motivation to lose weight, the further into her infertility journey she went, "Being disheartened doesn't inspire you to keep trying to lose the weight, it just makes you feel bad, like you don't even want to bother anymore. So, you just want to eat for the sake of eating because it's a complete waste of time trying to eat healthy."

"They don't want to go beyond what they see": Challenging Assumptions About Fatness and Infertility

Participants' accounts of the deleterious effects they experienced as a result of seeking support for infertility contradict the official discourse that the CPAC for fertility is a public health good because it inspires weight reduction (Farquhar & Gillett, 2006). We have traced how study participants internalised the problematisation of their weight in relation to fertility in harmful ways. However, some Indigenous and feminist critical scholars have argued that the process of subjectification to dominant medical discourses in women's health care is never total and point to the ways in which women both absorb and resist discourses that problematise their reproductive bodies (Le Grice, 2014; McKenzie-Mohr & Lafrance, 2014; Penehira et al., 2014). We identified a range of strategies used by participants to challenge their exclusion from fertility treatment, leading them to resist the conflation of fatness and infertility, assumptions about why people are fat, and the logic of weight reduction. Participants also imagined what just and compassionate fertility care would look like.

Despite the singular focus on their weight as both the cause of and solution to their infertility, participants questioned what other underlying causes might contribute to their infertility and felt frustrated and angry at the lack of investigation into them. Maia, who after being declined for public treatment had eventually managed to access treatment privately, became pregnant swiftly through a simple procedure and reflected angrily that her fatness had been assumed to be the issue:

> So, it was interesting that while I couldn't get funding to get any form of treatment or assistance, in actual fact it was a very simple initial procedure that led to my pregnancy. So, I was very, very upset that I'm a taxpayer, good citizen, you know, I tick all the boxes and the only thing that excluded me was the fact that I was fat.

Participants questioned why there was little or no interest in the health status (including the weight) of their partners, and why other issues they were experiencing, such as prolonged vaginal bleeding, were not investigated.

Participants also resisted the assumption that their fatness was a failure of self-management, giving various examples of how they enacted agency in health-seeking. Most participants rejected the assumption that their fatness was a result of inactivity and overeating 'bad' food, describing these assumptions as stereotyping and stigmatising. As Talia described:

> It's just very stereotyped, the way, especially doctors, they just view you and that's it, they don't want to go beyond what they see. They just see you as "ok, you're fat, let's sort that out first" rather than, "ok, well how about we look within and you would see that I'm actually a really healthy person apart from this."

Maia said: "It's almost like smoking. People think that you do it to yourself. People think that you have no control." Participants objected to a perceived lack of personal responsibility and agency to seek out healthy behaviours on their part and advocated for their weight to be viewed holistically in the context of their broader health. As Mere asked, "If you're a person that, say, eats healthy, is doing exercise on a daily basis, wouldn't you consider that and say "hey, we might be able to help you?"

Participants were also highly critical of the logic that weight loss was simply a matter of willpower that could be achieved if they were truly committed to becoming pregnant. Participants described the complexities of their lives, including shift work, family and childcare responsibilities, church and study commitments, and household responsibilities, all of which constrained their ability (in terms of both time and energy) to commit to a project of weight loss. As Mere reflected:

> Not everyone has the perfect hours and job to be able to manage a healthy lifestyle. And not everyone has got parents to help out with your child if you are trying to manage work around them. I do my best by my son first, and I try to not depend on family members.

Likewise, Natia described the impossible challenge of trying to lose weight while being a full-time mum, on rotation shift, and trying to study: "You know life just happens, like Uni, and I just found that all the late nights and assignments kind of really stressed me out, and I thought that I could do some exercise during that time but I just had to stay up too late."

Maia also pointed to normal human diversity in body sizes, which made substantial weight loss unrealistic for many people: "Like they wanted me to lose a considerable amount, like 30 to 40 kilos to be eligible and that was nigh on impossible, it just wasn't going to happen."

Finally, participants imagined a fairer, more just, and compassionate approach to fertility care. Participants wanted individualised, one-on-one assessment, and the development of a support plan to help them lose weight and/or identify and address the causes of their infertility. As Natia imagined: "More support, more kind of one-on-one care. I know they're busy but even if it's just having a plan for me. There's no plan. There's never been a plan." In the following section participants' ideas of improving fertility care are shown to synthesise with a reproductive justice approach and are proposed as a framework for improving fertility services for fat people.

Where to for Weight-Centric Fertility Care: A Reproductive Justice Agenda for Assisted Reproduction

Although Farquhar and Gillett (2006, p. 1109) claim that a BMI cut-off for fertility treatment access is "good medicine" because it promotes public health goals of weight reduction, the accounts of our participants tell a different story. Through an affective-discursive lens we have shown how being made to *feel* bad about their bodies in the process of seeking fertility care, and further being locked out of fertility care, was a damaging intervention in the lives of our participants, undermining their sense of self-worth, and worthiness as parents, harming their health, and leading to a loss of trust and faith in their doctors and the health care system at large. Further, keeping the doors of fertility services firmly closed until participants achieved weight loss did not work as an incentive, as is assumed in the medical literature, and for the majority of participants in this study presented an impossible task that resulted in further negative feelings of demoralisation and hopelessness. This was compounded by a lack of help or support with achieving these weight-reduction requirements. We suggest that these are unacceptable outcomes from reproductive health policy and defy public health logic. We note that, increasingly, there are challenges within fertility medicine that align with our findings (e.g., Legro, 2016): some studies are starting to question whether fertility outcomes are indeed better when weight loss is initiated prior to fertility treatment and suggest that women with high BMI should proceed directly into treatment (Legro, 2016). We note Legro's (2016, p. 2662) impassioned call, "Tear down that weight wall! Allow all obese women adequately informed of the risks and benefits of fertility treatment and subsequent pregnancy to proceed by their own free will with treatment."

Returning to the framework of reproductive justice, we problematise the use of personal responsibility and choice as the currency to determine who deserves fertility care. Reproductive justice critiques the notion of personal choice, focusing instead on the ways that reproductive decision-making can be understood in relation to the socio-political structures that shape and determine the choices people are able to make (Ross & Solinger, 2017). The emphasis on participants' personal responsibility for their weight denies the material realities of their lives. These material realities restrict their ability to make choices, both about weight loss and their ability to pursue fertility care outside of the public health system. Gendered differences in household and childcare responsibilities, socio-economic disparities that determine the type of paid work people do and how much they are paid for it, and raced differences that determine access to health care and health outcomes are all factors that shape our participants' choices in relation to weight loss and fertility, all of which are invisibilised by the emphasis on personal responsibility. For example, we note that of the two participants who were eventually able to resource private fertility treatment both fell pregnant with ease. Those who were not able to, due to socio-economic constraints, did not. Mere gave a stark articulation of this, observing that when you apply weight criteria across the board: "You pretty much cut out, what, 70 percent of the people in South Auckland, and where can we go, there's nowhere to go, other than paying for private and we don't have that money."

Further, weight-based criteria that do not attend to ethnic and Indigenous differences in physicality, and that deny existing inequalities experienced by Māori and Pacific people within and outside the healthcare system, result in the perpetuation and amplification of racist and colonial legacies of reproductive interference in the lives of Indigenous women and women of colour (e.g. Le Grice & Braun, 2016). We note Māori and Pacific participants' awareness of the role that fat shaming, racialisation, and socio-economic status have in shaping their access to publicly funded fertility treatment. Contrary to the notion that publicly funded fertility services are welcoming of all, participants viewed the CPAC criteria as an instrument of racialised exclusion. Our findings affirm and contribute to a growing critical literature that suggests obesity-related policies unfairly effect and exclude women along the lines of race, class, and physicality. Exclusion on these grounds risks being eugenicist in effect, if not intention (Bombak et al., 2016; LaMarre et al., 2020).

We conclude by once again drawing on the principles of reproductive justice to suggest that if the reproductive dignity, agency, and rights of all people seeking publicly funded fertility care are to be observed, we must insist on fertility care policy that decolonises reproductive technologies that support positive affectivities and promote fair, just, and inclusive access to fertility treatment. A reproductive justice agenda for assisted reproduction will necessarily abandon weight-based criteria and be grounded in a much more complex and socially just understanding of the relationship between body weight, reproductive health, and fertility (Parker et al., 2019). Such an approach would recognise that reproductive health and a positive pathway to parenthood can never be achieved by derogating those seeking fertility care and by denying the material realities of people's lives and circumstances (Friedman, 2015).

A reproductive justice paradigm for fertility care would insist on a holistic, compassionate, and arguably non-Western approach to fertility care that disrupts the universalisation of Western bodies and medical knowledges, attends to the social determinants of weight, fertility, and reproductive health, and acknowledges Indigenous and other cultural epistemologies of embodiment and reproduction (see Le Grice & Braun, 2016). The accounts of participants in our study reinforce the findings of LaMarre et al. (2020), which advocate for fertility care grounded in the principles of reproductive dignity and justice and make a plan for treatment that is tailored to reach their own health and fertility goals. Again, in the words of Natia: "More support, more kind of one-on-one care. I know they're busy but even if it's just having a plan for me. There's no plan. There's never been a plan."

Notes

1. In this chapter we draw on critical obesity and Fat Studies, using fat as a reclaimed word to describe bodies that are larger than those designated as 'normal' in Western culture and medical discourse and avoiding pathologising terms like obesity.
2. We acknowledge that not all gestational parents will identify as women or mothers, and not all parents who identify as women or mothers have, or will have, gestated children. We have followed the lead of Ross and Solinger (2017) by using both gendered terms such as "woman" and "mother," and gender-inclusive terms such as pregnant person/parents. Where participants were themselves woman-identified, we have referred to them as such. Where we refer to fertility services policy, we also use woman and mother. We have aimed to ensure our lexicon is gender-inclusive, but we acknowledge that the language used to describe gestational parent experience is imperfect and must continue to be challenged.
3. In Aotearoa, New Zealand, a general practitioner (GP) is a doctor who provides medical care in the community as part of a primary care practice and is usually the first point of contact with the healthcare system. GPs refer people to secondary- and tertiary-level healthcare as necessary.
4. Māori are Indigenous (tangata whenua) to Aotearoa New Zealand. Pasifika people are Indigenous to the islands of Te Moana Nui a Kiwa (The Great Ocean of Kiwa—The Pacific Ocean).
5. Auckland is Aotearoa New Zealand's largest city, with approximately 1.6 million inhabitants.
6. Intersectionality directs attention to the ways in which claims about women's experiences often produced 'hegemonic generalisations' by

universalising the experiences of and problems of privileged women (most often white, Western, middle-class, heterosexual women) to all women. Intersectional analysis seeks to complicate feminist analyses, ensuring attention to the "interlocking effects of identities, oppressions, and privileges" in order to fully understand the range and complexity of women's experiences (Price, 2011, p. 55).
7. Whānau is a Māori word, broadly translated in English to refer to extended family. Whakapapa is a fundamental principle in Māori culture that refers to genealogical links. It describes human connection and belonging through ancestry, incorporating past influences and future potential.

References

Armitage, J. A., Poston, L., & Taylor, P. D. (2008). Developmental origins of obesity and the metabolic syndrome: The role of maternal obesity. *Obesity and Metabolism, 36*, 73–84.

Bacchi, C. (2012). Why study problematizations? Making politics visible. *Open Journal of Political Science, 2*(1), 1–8.

Bacon, L., & Aphramor, L. (2011). Weight science: Evaluating the evidence for a paradigm shift. *Nutrition Journal, 10*(1), 9–30.

Bjerg, H., & Staunæs, D. (2011). Self-management through shame—Uniting governmentality studies and the 'affective turn'. *Ephemera: Theory & Politics in Organization, 11*(2), 138–156.

Bombak, A. E., McPhail, D., & Ward, P. (2016). Reproducing stigma: Interpreting "overweight" and "obese" women's experiences of weight-based discrimination in reproductive healthcare. *Social Science & Medicine, 166*, 94–101.

Braun, V., & Clarke, V. (2013). *Successful qualitative research: A practical guide for beginners*. Sage.

Chrisler, J. C., & Barney, A. (2017). Sizeism is a health hazard. *Fat Studies, 6*(1), 38–53.

Farquhar, C. M., & Gillett, W. R. (2006). Prioritising for fertility treatments—Should a high BMI exclude treatment? *BJOG: An International Journal of Obstetrics & Gynaecology, 113*(10), 1107–1109.

Friedman, M. (2015). Reproducing fat-phobia: Reproductive technologies and fat women's right to mother. *Journal of the Motherhood Initiative for Research and Community Involvement, 5*(2), 27–41.

Gard, M., & Wright, J. (2005). *The obesity epidemic: Science, morality and ideology*. Routledge.

Gavey, N. (2011). Feminist poststructuralism and discourse analysis revisited. *Psychology of Women Quarterly, 35*(1), 183–188.

Gillett, W., & Peek, J. (1997). Access to infertility services; development of priority criteria. Wellington: New Zealand National Health Committee. https://www.moh.govt.nz/notebook/nbbooks.nsf/0/EE485C7A94B3399E4C2565D70018B803/$file/Access-to-infertility-services.pdf

Gillett, W. R., Putt, T., & Farquhar, C. M. (2006). Prioritising for fertility treatments—The effect of excluding women with a high body mass index. *BJOG: An International Journal of Obstetrics & Gynaecology, 113*(10), 1218–1221.

Greil, A. L., Slauson-Blevins, K., & McQuillan, J. (2010). The experience of infertility: A review of recent literature. *Sociology of Health & Illness, 32*(1), 140–162.

Herdon, A. (2018). Overfeeding the floating fetus and future citizen: The "war on obesity" and the expansion of fetal rights. In J. Verseghy & S. Abel (Eds.), *Heavy burdens: Stories of motherhood and fatness* (pp. 35–44). Demeter.

LaMarre, A., Rice, C., Cook, K., & Friedman, M. (2020). Fat reproductive justice: Navigating the boundaries of reproductive health care. *Journal of Social Issues, 76*(12), 338–362.

Le Grice, J. (2014). *Māori and reproduction, sexuality education, maternity, and abortion* (PhD thesis). University of Auckland. https://researchspace.auckland.ac.nz/bitstream/handle/2292/23730/whole.pdf?sequence=2

Le Grice, J., & Braun, V. (2016). Mātauranga Māori and reproduction: Inscribing connections between the natural environment, kin and the body. *AlterNative: An International Journal of Indigenous Peoples, 12*(2), 151–164.

LeBesco, K. (2011). Neoliberalism, public health, and the moral perils of fatness. *Critical Public Health, 21*(2), 153–164.

Legro, R. S. (2016). Mr. Fertility authority, tear down that weight wall! *Human Reproduction, 31*(12), 2662–2664.

McKenzie-Mohr, S., & Lafrance, M. N. (Eds.). (2014). *Women voicing resistance: Discursive and narrative explorations*. Routledge.

Ministry of Health. (2019). Annual update of key results 2018/19: New Zealand Health Survey. Retrieved November 17, 2020, from https://www.health.govt.nz/publication/annual-update-key-results-2018-19-new-zealand-health-survey

Ministry of Health. (2020). *Obesity statistics*. Retrieved November 17, 2020, from http://www.health.govt.nz/nz-health-statistics/health-statistics-and-data-sets/obesity-statistics

National Women's Health. (2020). *Public funding*. Retrieved November 11, 2020, from https://www.nationalwomenshealth.adhb.govt.nz/our-services/fertility/publicfunding/#:~:text=Woman's%20BMI%20must%20be%20less,than%20or%20equal%20to%2032

O'Hara, L., & Gregg, J. (2006). The war on obesity: A social determinant of health. *Health Promotion Journal of Australia, 17*(3), 260–263.

Pandey, S., Maheshwari, A., & Bhattacharya, S. (2010). Should access to fertility treatment be determined by female body mass index? *Human Reproduction, 25*(4), 815–820.

Parker, G. (2014). Mothers at large: Responsibilizing the pregnant self for the "obesity epidemic". *Fat Studies, 3*(2), 101–118.

Parker, G., & Pausé, C. (2019). Productive but not constructive: The work of shame in the affective governance of fat pregnancy. *Feminism & Psychology, 29*(2), 250–268.

Parker, G., Pausé, C., & Le Grice, J. (2019). "You're just another friggin' number to add to the problem": Constructing the racialised (m)other in contemporary discourses of pregnancy fatness. In M. Friedman, C. Rice, & J. Rinaldi (Eds.), *Thickening fat: Fat bodies, intersectionality and social justice* (pp. 97–109). Routledge.

Penehira, M., Green, A., Smith, L. T., & Aspin, C. (2014). Māori and indigenous views on R and R: Resistance and resilience. *MAI Journal, 3*(2), 96–110.

Poston, L., Caleyachetty, R., Cnattingius, S., Corvalán, C., Uauy, R., Herring, S., & Gillman, M. W. (2016). Preconceptional and maternal obesity: Epidemiology and health consequences. *The Lancet Diabetes & Endocrinology, 4*(12), 1025–1036.

Price, K. (2011). It's not just about abortion: Incorporating intersectionality in research about women of color and reproduction. *Women's Health Issues, 21*(3), 55–57.

Ross, L., & Solinger, R. (2017). *Reproductive justice: An introduction*. University of California Press.

Sanders, R. (2019). The color of fat: Racializing obesity, recuperating whiteness, and reproducing injustice. *Politics, Groups, and Identities, 7*(2), 287–304.

Ward, P., & McPhail, D. (2019). Fat shame and blame in reproductive care: Implications for ethical health care interactions. *Women's Reproductive Health, 6*(4), 225–241.

Ware, F., Breheny, M., & Forster, F. (2017). The politics of government 'support' in Aotearoa/New Zealand: Reinforcing and reproducing the poor citizenship of young Māori parents. *Critical Social Policy, 37*(4), 499–519.

Wetherell, M., McCreanor, T., McConville, A., Barnes, H. M., & Le Grice, J. (2015). Settling space and covering the nation: Some conceptual considerations in analysing affect and discourse. *Emotion, Space and Society, 16*, 56–64.

8

The Experience of Single Mothers by Choice Making Early Contact with Open-Identity or Private Sperm Donors and/or Donor Sibling Families in New Zealand

Rochelle Trail and Sonja Goedeke

The term 'Single Mothers by Choice,' or 'Choice Mothers,' refers to women who opt to become single mothers through sperm donor insemination or adoption (Jadva et al., 2009a). Use of the term dates back to 1981, when Jane Mattes, an American psychotherapist and single mother by choice herself founded the 'Single Mother by Choice' organisation (www.singlemothersbychoice.org). In 2003, Mikki Morrissette founded the 'Choice Moms' website (www.choicemoms.org) to reflect that while being single is not necessarily a choice, becoming a parent while single may be a conscious rather than a circumstantial decision. While these terms can be used interchangeably, in this chapter we adopt the term 'Choice Mothers' to describe women in the New Zealand context who actively choose to embark on parenthood alone.

R. Trail (✉) • S. Goedeke
Auckland University of Technology, Auckland, New Zealand
e-mail: sonja.goedeke@aut.ac.nz

As a sub-group of single parents, Choice Mothers may be viewed as challenging dominant paradigms around reproduction and family-building, as reflected by traditional nuclear family households (Ajandi, 2011). In previous research conducted on the experience of being a Choice Mother in New Zealand (Trail & Goedeke, 2019), we noted that most of our participants were able to recall at least one incident where they had been on the receiving end of a less than positive reaction to their decision to be a Choice Mother. Choice Mothers occupy a socially marginalised position in relation to their family-building and may need to disregard possible negative public perception to pursue single motherhood. As research by Susan Boyd suggests, 'for women to undertake single parenthood in the light of still powerful negative discourses takes courage' (Boyd, 2019, p. 94).

The majority of women who become Choice Mothers do so with the assistance of donor insemination (Jadva et al., 2009a), as in New Zealand, where the numbers of single women seeking fertility treatment is on the rise. At the largest fertility clinic in New Zealand, Fertility Associates, the numbers of single women treated with donor insemination almost doubled over a three-year period, from 80 in 2012, up to 156 in 2015 (Bilby, 2015), and by 2018, Choice Mothers made up around 56% of all donor insemination treatments (J. Weren, personal communication, October 26, 2018).

Historically, sperm donor insemination through fertility clinics in New Zealand was available to heterosexual couples only (New Zealand Law Commission, n.d.), but following a complaint lodged with the Human Rights Commission in 1993, fertility clinics in New Zealand have allowed single women and those in same gender relationships to seek treatment (New Zealand Law Commission, n.d.). Sperm donors are, however, given a choice in terms of who is able to use their sperm and may choose to place restrictions on donations, such as excluding single women (Advisory Committee on Assisted Reproductive Technology, 2015), which implies that single women are likely to wait longer to access sperm than women in heterosexual relationships.

In New Zealand, the Human Assisted Reproductive Technology (HART) Act of 2004 recognises the rights of donor offspring to be made aware of their genetic origins. Men who donate sperm through clinics are required to be identifiable with their details recorded on the HART register and available to donor offspring when they reach 18 or sooner by application (usually at 16). Recipients of gamete donation may also request identifying information from the clinic about the donor when their child is younger than 18 years old. Further information about donor siblings may be requested and released if the donor sibling is over 18 and has given permission to do so, or if under 18, with their parents' consent (Fertility Associates, n.d.).

In addition to clinic 'open-identity' donors, options for single women are to access donors through alternative sources, such as approaching friends and acquaintances or by seeking donors online. In these instances, conception may take place via artificial insemination and without the involvement of a fertility clinic. 'Private donors' (terminology we use for the purposes of this chapter) refers to donors recruited by single mothers themselves for the sole purposes of conception. Note that in this case there is no obligation for any third party to hold information on the donor, which means it is possible that donors' details become lost over time if contact is not maintained or if conflict arises.

New Zealand's open legislation and practice reflects that of many other countries such as Sweden, Norway, United Kingdom, the Netherlands, and Australia in the removal of donor anonymity and the recognition of the importance of access to genetic knowledge for healthy identity formation (Daniels & Douglass, 2008). This is given particular credence in New Zealand, where the beliefs of Indigenous Māori about the importance of genealogy (whakapapa) are crucial in locating individuals within broader networks of kin relations and belonging (Shaw, 2008; Webber, n.d., p. 17). Thus, the Māori cultural perspective has had a critical influence on the New Zealand approach to open-identity donation (Daniels, 2007) in recognising the importance of kinship and understanding the necessity of donor-conceived people to locate their 'place in the world' (Daniels & Douglass, 2008).

While there is a link between using open-identity donors and disclosure to offspring about the nature of their conception, the latter does not necessarily appear to follow the former. Some research has found that despite a move towards intentions of increased transparency, many recipients still hesitate to disclose (Nordqvist, 2014; Readings et al., 2011).

This is in spite of research that indicates the importance of revealing the donor conception to the offspring's subsequent wellbeing (Cushing, 2010; Mahlstedt et al., 2010; Turner & Coyle, 2000) and the negative consequences of later disclosure (Golombok et al., 2013; Scheib et al., 2003; Turner & Coyle, 2000). Early disclosure has been associated with a greater ability by offspring to integrate the nature of their conception into their identity (Persaud et al., 2017).

For Choice Mothers, the situation is clearly somewhat different from that of heterosexual couples as the obvious lack of a father needs to be acknowledged and explained to offspring. Indeed, studies show that Choice Mothers are positive about their decision to use identifiable donors, believe their child's genetic origins to be of importance to them (Freeman et al., 2016), and are more likely to tell their offspring about their donor conception and at an earlier age (Beeson et al., 2011; Jadva et al., 2010).

Research also suggests that the majority of donor offspring, regardless of their family composition, desire contact with their donor (Beeson et al., 2011). Motivations include the need to understand medical history; to avoid consanguineous relationships; to satisfy curiosity; to learn more about self or complete one's identity; to learn more about the donor as a person; and to learn about ancestry/genealogy (Ravelingien et al., 2013). Ravelingien et al. (2013) found that access to the donor's identity alone is not sufficient for many of these reasons, and that some offspring wish to have contact or even form a relationship with the donor. This finding has been backed up by more recent research by Kelly et al. (2019b) in the Victorian state of Australia, where it was found that donor-conceived adults desired ongoing contact with their donor. Interestingly, other research has found that being the child of a Choice Mother is related to an increased likelihood of seeking information or donor contact (Scheib et al., 2017).

Similarly, research has also found that most donors are open to some kind of future contact with offspring (Daniels et al., 2012; Kelly et al., 2019b). Studies also indicate that donor siblings may be eager to establish contact (Jadva et al., 2010; Jadva et al., 2009b) and that some gamete recipients, especially in the case of Choice Mothers or lesbian mothers, may be keen to initiate contact on behalf of their children (Freeman et al., 2009; Kelly & Dempsey, 2016a; Scheib & Ruby, 2008).

Donor-linking may take place via a statutory linking regime, via 'informal means' such as online registries (e.g., www.donorsiblingregistry.com) or directly through fertility clinics (Kelly & Dempsey, 2016b). Kelly et al. (2019a), however, in their research on donor-linking by 12 fertility clinics across Australia, identified an ad hoc approach where clinics may either take an active or passive approach to donor-linking, resulting in vastly different experiences and access to information. To date, there is little research on the donor-linking experience in New Zealand.

The state of Victoria, in Australia, has one of the most comprehensive donor-linking legislative frameworks in the world and a statutory authority, the Victorian Assisted Reproductive Treatment Authority (VARTA), was established to manage donor-linking and provide support and free counselling to applicants in this state (Dempsey et al., 2019). Legislation was also enacted retrospectively even giving applicants who were conceived or donated during the period of anonymity, prior to 1988, the ability to request identifying information, thus opening the possibility of contact (Kelly et al., 2019b). Research on the impact of such contact is emerging (Crawshaw et al., 2015; Dempsey et al., 2019; Kelly & Dempsey, 2016a; Kelly et al., 2019a). In Dempsey et al.'s (2019) study, the largest group of applicants to apply to VARTA for personal information and contact were recipient parents and in particular, Choice Mothers. It was found that motivations for contact included, 'to have information available to their child in the future; because the child had asked for information; and to explain observed aspects of their child's personality' (Dempsey et al., 2019, p. 32).

To date, research suggests that such contact may have positive consequences for offspring, donors, and donor siblings, through satisfying some curiosity about their donor and family origins and forming extended family networks (Persaud et al., 2017; Scheib & Ruby, 2008). However, challenges have also been identified such as the need to establish limits and boundaries regarding contact (Daniels et al., 2012; Hertz et al., 2015), and to manage the effects on relationships with donors' partners, their children, and possibly extended family members. Such challenges raise the question of the provision of appropriate support for those seeking to build their families through donor conception, and as single women.

Current research on the experience of contact between donors tends to be confounded by various factors. This includes research drawing on data from anonymous donors or of different family forms and occurring in contexts with different legislative frameworks regarding contact, including the age at which contact may be possible. While Choice Mothers have been reported to be more likely to seek early contact with donors as well as donor siblings (Kelly & Dempsey, 2016b), how such contact is experienced and what support Choice Mothers and their offspring may need, are thus important to consider.

Study Methods

In this chapter, we draw on the findings of a study by the first author, in which six Choice Mothers were interviewed as to their experiences in the New Zealand context. The research took place in the latter half of 2018 with the participants interviewed over a period of six weeks between July and August 2018, either face-to-face or via Skype. Most were based in Auckland with one participant living in a small town outside of Auckland. Participants were aged between 33 and 49 years of age and 5 identified as Pākehā or European New Zealanders, 1 as being of mixed ethnicity. All women had only one child, although one woman was in a relationship with step-children. Participants' children were aged between three months and seven years old at the time of the research.

All women were interviewed using a semi-structured interview format as to their experiences of seeking or making early contact with 'open-identity' or 'private donors.' Data from the interviews was transcribed and analysed using Braun and Clarke's (2006) well-established six-step method of data analysis. This included: data familiarisation, assigning preliminary codes, searching for patterns across the interviews, revision of themes, defining and naming of themes, and finally, producing a report of the data.

The small sample size reflects some of the challenges in recruiting within a relatively small, but growing community of Choice Mothers.

While a small sample size limits the degree of extrapolation across a wider community, it was deemed that the participants were reasonably representative of the women that choose to become Choice Mothers who have been described as older, professional, and financially secure (Murray & Golombok, 2005; Jadva et al., 2009a).

Our discussion focuses on the motivations of these women in seeking early contact, their intentions and experiences in disclosing donor conception to their offspring and negotiating contact, and the impact on constructs of family. We outline how for these women disclosure to their offspring is perceived as ongoing, rather than a one-off event; how contact is constructed as more beneficial than identifiability alone; how expectations and boundaries present challenges to establishing contact; and how family constructs are shaped in the context of open identity and contact.

The Importance of Donor: Offspring Contact

New Zealand has a long history of advocating for openness and transparency (Daniels & Douglass, 2008), and as described above, although legislation to prohibit the anonymity of donors was only formally enacted in the HART Act 2004 (Fertility Associates, n.d.), open practice has been common in New Zealand at least since the 1990s (Shaw, 2008). New Zealand has a relatively unique approach in contrast to most international practices, in that the donor's identifying information may be available to *recipients* by request once their child is born (Fertility Associates, n.d.). Most jurisdictions that offer open-identity donors require information to be made available to the *offspring* only at the age of majority or younger by application. Early contact is thus possible in New Zealand not only for private donors, but also clinic-recruited donors, and was found to be desired by the Choice Mothers in our study.

In New Zealand, legislation allows for such contact between parties to be facilitated either through Births, Deaths, and Marriages (BDM) or through the fertility clinics. It is preferred that contact be facilitated through the clinics so that counselling can be conducted, preferably with both parties, before their first face-to-face meeting takes place (J. Peek, personal communication, November 10, 2018). Fertility clinics in New Zealand explain to donors that they should be open to contact prior to the offspring becoming 18, in part due to adolescence being regarded as a key period for identity formation (Erikson, 1959). Efforts to facilitate this process could thus be envisaged to take place earlier in the child's life than the current age of maturity at 18 years.

While our study consisted of a small sample size, all six Choice Mothers had initiated contact with the donor soon after their child was born and emphasised its importance. This interest in early contact was also reported in Kelly and Dempsey's (2016b) study, which is relatively unique in that it investigates situations where the recipient, and not the offspring, sought early contact. A more recent Australian study found that some of the recipient parents requesting donor-linking also had children as young as two years old (Kelly et al., 2019a).

All the women in our study regarded early contact as opening the way for a connection between the donor and their offspring and as a significant step in assisting with their child's healthy identity formation. For example, Natasha said: 'I think being able to know your place in the world and understand how you came into the world is very important. I think being able to trace that is hugely important (to) understand your identity.'

To this end, these women had specifically selected donors who had agreed to be contactable (prior to the offspring reaching 18), indicating that it was contact with the donor, rather than identifying information, which was considered most important. They saw direct contact as providing for a better opportunity for their child to 'fill in some of the missing pieces,' 'make sense of their place in the world,' and for some, they related the importance of understanding their lineage and where they came from, back to their own identity struggles during adolescence. Further, Choice Mothers reflected that they wanted their child to be able to potentially form some kind of relationship with the donor and possibly other donor siblings. Millie commented:

> '... the best situation for me, would be for them to have a relationship, be friends... that would be the perfect scenario, like so they all know each

other, and you know they can pick up the phone and ask questions to each other at times.'

In both cases where a private donor had been used, contact continued with the donor once their child was born. These findings mirror other studies that suggest that contact and connection with the donor, rather than access to identifying information, may be desired, partly as a means to develop a greater sense of identity but also to build relationships (Cushing, 2010; Dempsey et al., 2019; Kelly et al., 2019b; Ravelingien et al., 2015).

Initial contact in our study was facilitated through the fertility clinics in the four cases where women had utilised the clinic to conceive, with staff generally acting as a 'go-between' by passing correspondence between parties. However, despite clinic guidelines, only one of the participants had received counselling with regard to contact with the donor. Furthermore, while the clinics were the main source of contact between the parties, Choice Mothers expressed concern about the way in which communication was facilitated and information managed, and a need for more specific support. Once an initial connection had been made, how much contact to have, and what form that might take (i.e., direct email contact, photographs, or a face-to-face interaction) was at the time of our study often still being negotiated by Choice Mothers. Some participants reflected that once they had made an initial connection, they were comfortable with not seeking further contact until their children expressed a desire to meet the donor face-to-face, which they envisaged could be when their children began asking and confronting more identity-related questions. Cassidy reflected:

> 'So, I guess you know, it's up to [son], he might be curious, or he might be just like actually, this isn't my bag, and I'm happy with that. I'll be led by whatever he wants to do because it's not my decision, I guess.'

All the women were united, however, in believing that having the opportunity for contact was an important consideration in the formation of their child's sense of their identity. With the exception of one participant, Choice Mothers highlighted that they would likely have rejected the option of an anonymous donor if they had had to seek overseas treatment where anonymous donation options may be more common.

The Need for Clear Expectations and Boundaries in Donor: Offspring Contact

Managing contact between donor and recipient/offspring in the absence of clear guidelines and support to facilitate this process was regarded as very challenging and as navigating new frontiers for both parties. In part, this stemmed for a need to clarify motivations for contact as these also influenced the expectations each party had of the other. Similar to Kelly and Dempsey's (2016b) research, for the Choice Mothers in our study, early contact was envisaged as an opportunity to thank the donor as well as an attempt to ensure that the donor was aware of the existence of the donor-conceived child and their wellbeing. It was regarded as paving the way for appropriate future contact between the donor and offspring. Choice Mothers wanted donors to be interested in the wellbeing of their offspring and desired some level of contact, but also wanted to negotiate and manage contact on their terms. This presented several potential challenges for the Choice Mothers in our study, including the donor not being the type of person they had imagined, the donor not finding them or their offspring to be what he in turn, had expected, and different ideas around contact. Natasha commented on situations where other women she knew had met with their donor:

> 'And maybe that's because (the clinic) is painting out this picture of this slightly romantic connotation… or… that they (the women) wish they hadn't met them (donors) because maybe they had a different idea, and then this person seemed very unusual. And of course, then they're left thinking, 'what does that mean for my child'?

In particular, the narratives that Choice Mothers had about their donor and who they perceived him to be set up expectations, which could create challenges around boundaries. From the outset of the donation process, Choice Mothers in our study had attributed admirable qualities to the donor, framing him as being a 'kind' and 'great guy' based upon his motivations to donate. They also viewed him as a crucial part of their child's identity and had built up a narrative of how a relationship between them and the offspring might play out. To a certain extent, the fertility clinic, beginning with the process of donor selection, promoted such narratives. Indeed, some research has suggested that fertility clinics actively manipulate donor profiles with the intention of enhancing the recipients' emotional connection with donors in order to remove the sense of donation being a commercial transaction (Bokek-Cohen & Gonen, 2015). It would seem that there is some need by both recipients and offspring (Turner & Coyle, 2000) to fantasise about the nature of the sperm donor in a positive way, especially in the absence of any prior direct contact with the donor. Such positive narratives about the donors, however, helped contribute to a desire for early contact, as well as a sense of expectation about how this contact would unfold.

Two of the six Choice Mothers in our study expressed some disappointment as to how their initial contact was received by the donors. For example, Cassidy said:

> 'I wasn't expecting lots of communication but just kind of, yeah, letters and you know, if he wanted photos or whatever, but I guess his response—I felt like he still wanted to keep it at arms-length.'

Both had to reconcile their ideal scenario with the reality of the situation. In one case, while the donor (and his partner) was happy to meet the recipient, they had expressed the desire not to meet her son at that time, which she experienced as quite challenging. The other Choice Mother had felt somewhat rebuffed by her donor's response to her initial contact and was apprehensive of pursuing contact, fearing that she might come across like a 'stalker,' a finding reported in research elsewhere (Kelly & Dempsey, 2016a). It is possible that opportunities to establish a relationship were compromised due to the narratives that had been constructed about the donor, and unclear expectations as a result.

Choice Mothers also spoke of perceived mixed messages from fertility staff members in relation to contact between donors and recipients. While clinics offer support to facilitate contact, Choice Mothers felt that they were given clear underlying messages that they should not make any demands on the donor, given his 'gift' of donation. With regard to Natasha, the Choice Mother whose donor was not ready to meet her son initially, she spoke of how she was counselled by the clinic to manage her expectations and felt chastised for pursuing a relationship with the donor. She said,

> 'I don't think they (the clinic) want us prying… I think they want to put us off so that we leave the poor donors alone… They've donated. They've helped you with a baby. Now leave them alone. Why do you need to make contact with them? That sort of thing.'

While she attempted a second pregnancy using his sperm (which was ultimately unsuccessful), she chose not to pursue further contact and distanced herself from him, both for fear that he would not live up to the fantasy she had constructed of him, and that he would not be open to ongoing connection.

Of particular interest to this research are the boundary issues that may arise when the recipient is a Choice Mother. Other research has identified the complexities of managing relationships with the donor and their networks. Of specific concern to the donor's partner can be the relationship of the donor both to donor offspring, but also their mother, particularly where the mother is a single woman (Daniels et al., 2012). In order to mitigate some of the potential for discomfort around this situation, one of the women with a private donor ensured that the majority of her contact was with the sperm donor's wife. Her terminology also reflected this by referring to them as 'the sperm parents,' thus choosing an inclusive title for both the donor and his wife.

This choice of terminology reflects the difficulty which the various parties may have in terms of defining the relationship between the recipient, the donor, the offspring, and the donor's extended networks (their own family unit), and the challenges to constructs of what constitutes a family. These ambiguities translated into Choice Mothers' unclear expectations of the donor and struggling to know how to define him and his relationship to their family. While Choice Mothers did not necessarily regard the donor as part of their family, they struggled to construct a narrative of his presence (or absence) in their family and his role remained vague and undefined. This was sometimes in contrast to the view they tended to hold of donor siblings who were often seen as extended family members. However, Erin had a more encompassing view of family and spoke of being thrilled to have received a card from her donor and his family and what that might mean for future contact: 'yeah, and I thought wow that is really cool cos… if she (my daughter) wants to meet her extended whānau (family) in the future… it's so reassuring.'

Issues may also arise on the donors' side which may influence contact expectations and arrangements. For example, in the study by Daniels et al. (2012), donors expressed some anxiety around connection especially in relation to being rejected or perceived as a disappointment to the recipient and offspring. This finding was mirrored in research by Isaksson et al. (2014) who further report that the reasons up to a third of donors do not want information about offspring from the clinic were related to a fear of disappointment if the offspring chose not to seek contact with them at a later date. These findings are interesting in that donors' reluctance to establish future contact may not necessarily be related to a lack of interest in connection, but rather, as a result of challenging and conflicting emotions.

The implications of this are that both parties may be left in limbo with neither really understanding what the other wants from this initial contact. Both may be engaged in an attempt to protect themselves from complicated emotions associated with the ambiguity of the situation. Choice Mothers may be fearful of being seen as 'stalkers,' as above, and as being overly eager in pursuing contact. This is in spite of research which suggests that sperm donors believe that they should not be the ones to initiate contact but that the type and frequency of contact should be determined by the needs of the recipients or the offspring (Daniels et al., 2012). This is an issue which needs to be addressed since, as highlighted in the New Issues in Legal Parenthood report (2005) in New Zealand, the majority of sperm donors do appear to be open to being contacted. On the other hand, they also do not want to receive 'a knock at the door' and want contact to be carefully managed and to be facilitated by clinics (New Zealand Law Commission, 2005). Similarly, Daniels et al. (2012) reported, most donors regard boundary setting as important to protect all parties. In New Zealand the HART Act provides protection for donors from either custodial rights or financial obligations towards the offspring, and legislation probably provides sufficient protection for private and known donors, although this is less clear. Following the recommendations of the Law Society report, allowing private and known donors' protection under the HART Act seems like the best option to remove some of the potential legal uncertainty faced in these situations (New Zealand Law Commission, 2005).

In sum, the need to clarify expectations of each other from the outset, including in relation to contact, as well as establishing clear boundaries, were identified as key issues in our study. A body akin to VARTA, which can offer donor-linking services and an array of options such as support and counselling for all parties in navigating this path, appears to be warranted. An independent organisation would remove the burden of providing this service from clinics which appear to be under-resourced and potentially commercially conflicted in this area.

Flexible Family Constructs

In seeking contact with their donor, conventional family constructs may be challenged, as reflected by the broad and flexible constructs of family held by the Choice Mothers in our study. Cassidy, who lives with her partner and his children, and her son, for example, commented:

> 'And even now we're in a blended family. The kids… see (my son) as their brother but they're obviously not related by blood. And then I've got my parents who are not related to these kids but I'm like, family is family.'

Many of the participants felt that their family included their extended family, often their parents and their siblings, and in some cases donor sibling families as well. The amount of contact with donor siblings varied from one participant who expressed little interest in donor siblings to others who were in regular contact with donor sibling families.

Furthermore, Choice Mothers spoke of how their family-of-origin formation, which frequently appeared to involve a separation of biological and social ties, had often informed and given them insight into alternative family-building options and had enabled them to embrace a sense of fluidity around family constructs. While recognising genetic lineage as important in identity formation, the majority of the participants in this research saw biological relatedness as less important in family construction than social connections. They spoke of the importance of the normalisation of all family forms—ones that were not necessarily driven by blood ties but often by emotional ties or by a combination of genetic, gestational, and social ties. For example, one Choice Mother using a private donor spoke of her 'donor family,' consisting of the donor, his wife, and their child. This framing also reflected her need to manage boundaries around including her donor's wife—and not just the donor himself, as an important part of her family.

The imperative to know one's genetic origins has been described in some research as taking a heteronormative approach to family construction, as it implies that something may be 'missing' (Michelle, 2006). Thus, for Choice Mothers there is a tension around being perceived as a sufficient family unit despite the absence of a father-figure and acknowledging the importance of genetic information and remedying the absence of a physically present 'father' by ensuring that there is at least access to information about the donor, if not to the donor himself. This concept forms part of the delicate balance around normalising alternative family formations while acknowledging their potential complexities.

Most of the Choice Mothers in our study had accepted the absence of a father-figure in their family and the possible negative connotations that may be associated with that choice through their happiness at the family they had created for themselves. While none of the women appeared to be looking for the donor to fulfil the role of the 'father' in their offspring's life, the donor's possible role in their family remained relatively vague with no real clarity on how to refer to him. Interestingly, Shelby, the Choice Mother who referred to the donor (and his wife) as 'the sperm parents,' appeared to acknowledge the donor and his wife as part of her family unit; however, the same participant made a clear distinction between genetic and emotional ties when referring to her daughter's donor siblings, with the effect of distancing them from the family. Shelby stated: 'But some people say, oh, they're siblings and I'm like, no, they're not. Genetically related, but to me a sibling is because of an emotional relationship. If that makes sense.'

This was in contrast with the other Choice Mothers in the study, who were able to embrace and name donor sibling relationships more easily than the relationship with the donor himself. Indeed, most of the Choice Mothers considered relationships with donor sibling families to be especially important due to the fact that none of the children in this study had full biological siblings. For the most part, donor siblings were embraced as full family members, and were viewed as helping their children to feel less alone in the world.

Indeed, research has indicated that relationships with donor siblings can be a rewarding experience and can positively influence identity formation (Blyth, 2012; Persaud et al., 2017). It has been described in some research as 'creating new hybridized family forms which reimagine traditional notions of kinship' (Bailey, 2018, p. 199). Relationships with the donor sibling families extends the complexities associated with donor: recipient/offspring contact, as high expectations may exist as Choice Mothers seek to extend the family construct beyond just the two of them and as they incorporate donor siblings into their construct of their family.

Disclosure as an Ongoing and Challenging Process

Much recent research has been in favour of disclosure of donor conception for reasons including that secrecy in relationships is often detected and that when children find out inadvertently much damage may be caused through a sense of mistrust of and betrayal from their parents (Daniels, 2007; Golombok et al., 2013; Turner & Coyle, 2000). In New Zealand, an implication of the 'open' legislation is that donors may apply to BDM to access identifying information with regard to any adult offspring that have been born as a result of their donation. Furthermore, donor siblings may also apply to access information about other donor offspring. While such offspring/siblings have the right to refuse this request, this attempt at contact may be the first time they are aware that they are donor-conceived if disclosure by their parents has not occurred. Fertility Associates are currently in discussions with BDM regarding this issue (J. Peek, personal communication, November 10, 2018). Nonetheless, these are pragmatic factors which, alongside moral imperatives, may encourage disclosure. This issue was highlighted when the state of Victoria in Australia enacted retrospective legislation allowing donors to seek contact with offspring. It was hoped that this legislation would encourage recipient parents to disclose the nature of conception to their offspring; however, this has not always been the case. There has been some publicity over at least two cases of donor-conceived offspring discovering the nature of their conception via donor applications through VARTA (Kelly et al., 2019b).

In our research, Choice Mothers expressed a desire to be open and to disclose the conception to their offspring. This is in keeping with research which suggests that single women tend to disclose more frequently and earlier (cited in Persaud et al., 2017). On the other hand, as in Freeman et al.'s (2016) study, this may not always be the case and it is important not to assume that the absence of a father may cause single women to disclose earlier, or at all. For example, one of the Choice Mothers in our study who was now in a relationship appeared to struggle more with disclosure, perhaps as her situation no longer required her to explain father absence as her son had taken to calling her partner 'dad.' Cassidy said,

> 'I'm sure one day he'll say, "who's my dad'? And then I will obviously have that conversation… But part of it, they're a kid, and they should be kids and they don't need to know about adult stuff until they're kind of ready for it, you know?'

Choice Mother families in these circumstances may become subject to the same challenges that face other families using donor insemination to have their families, including the possible fears around the impact of the relationship between the offspring and the father-figure. Additionally, this also adds to the complexity of how to explain donor conception within the wider family context including step-siblings who may not yet understand the 'facts of life.' There is also the need to consider ongoing disclosure discussions based on the developmental needs of the child and of changing family circumstances.

In some cases a failure to disclose may be related to a lack of knowing how to disclose. While most of the women in our study spoke of talking openly with their child about their conception, they also spoke of challenges of disclosure, including how to talk about the 'facts of life' and having access to information, such as age-appropriate explanation for their offspring and for their children's peers. Natasha reflected:

> 'We've talked about it a little bit because he was struggling to communicate with other kids. And so, I've tried to give him some words that would help him say it clearly… So, from that discussion we started talking a little bit more about the words so that he could feel comfortable.'

Some New Zealand–based research has shown that non-disclosure has been directly linked with a lack of targeted counselling from clinics, specifically in the process of disclosure (Hargreaves & Daniels, 2007). This indicates that support in this area is likely to assist many parents including Choice Mothers. As children's needs and that of their peers change over time, it is important that disclosure is seen as a process rather than an event (MacDougall et al., 2007) and is managed by using a 'scaffolding approach' or in the form of workshops (Crawshaw & Montuschi, 2013) such as those offered by VARTA, that allow ample time to explore ideas.

Conclusion

While Choice Mothers may emphasise social, rather than genetic, ties in the building of their families, they also recognise lineage and genetics as related to identity, and the potential role of donor-offspring contact in facilitating healthy identity development for donor-conceived individuals. Choice Mothers may thus value early contact with donors in an attempt to pave the way for future contact between the donor and offspring. Contact may also be seen as a way in which to create an extended family, especially when contact includes donor siblings, and may be considered important given the smaller family size of most Choice Mothers. Early contact may also present a number of challenges, however, especially when Choice Mothers hold a romanticised narrative of the donor as a person and his desire for contact. Challenges may include negotiating boundaries with regard to initiating and maintaining contact in a mutually satisfying and safe way. Choice Mothers may need support to facilitate disclosure to offspring and in managing expectations and boundaries around contact. Some outstanding recommendations by the Law Commission, such as ensuring that recipients of donor gametes attend educational programmes to understand fully the implications of using open-identity donors, recommendations around disclosure, and options for making contact, may need further consideration to support the wellbeing of all parties involved in this form of family-building. While contact and relationships between parties cannot be legislated or enforced, early contact may be desired by some Choice Mothers and may be deemed beneficial for offspring. It would seem more focus is warranted to ensure that this process is managed well to ensure the best possible outcome for all concerned, not least that of donor-conceived offspring (see Chap. 12 in this volume for a discussion). And for this reason, it is paramount that the voices of donor-conceived people themselves are heard in order to determine the processes and legislation that directly impact their future options.

References

Advisory Committee on Assisted Reproductive Technology. (2015). *Informed consent and assisted reproductive technology: Proposed advice to the Minister of Health*. Advisory Committee on Assisted Reproductive Technology.

Ajandi, J. (2011). Single mothers by Choice: Disrupting dominant discourses of the family through social justice alternatives. *International Journal of Child, Youth and Family Studies, 2*(3/4), 410–431.

Bailey, E. (2018). Writing new branches into being: Connecting donor-linked families via Web 2.0. *Interactions: Studies in Communication & Culture, 9*(2), 195–206.

Beeson, D., Jennings, P., & Kramer, W. (2011). Offspring searching for their sperm donors: How family type shapes the process. *Human Reproduction, 26*(9), 2415–2424. https://doi.org/10.1093/humrep/der202

Bilby, L. (2015, May 10). More career-minded single women seek IVF treatments. *NZ Herald.* https://www.nzherald.co.nz

Blyth, E. (2012). Genes r us? Making sense of genetic and non-genetic relationships following anonymous donor insemination. *Reproductive biomedicine online, 24*(7), 719–726.

Bokek-Cohen, Y. A., & Gonen, L. D. (2015). Sperm and simulacra: Emotional capitalism and sperm donation industry. *New Genetics and Society, 34*(3), 243–273.

Boyd, S. B. (2019). Choice and constraint: Exploring 'Autonomous Motherhood'. In H. Willekens, K. Scheiwe, T. Richarz, & E. Schumann (Eds.), *Motherhood and the law* (pp. 73–100). University of Göttingen.

Braun, V., & Clarke, V. (2006). Using thematic analysis in psychology. *Qualitative Research in Psychology, 3*(2), 77–101.

Crawshaw, M., Daniels, K., Adams, D., Bourne, K., van Hooff, J. A. P., Kramer, W., … Thorn, P. (2015). Emerging models for facilitating contact between people genetically related through donor conception: A preliminary analysis and discussion. *Reproductive Biomedicine & Society Online, 1*(2), 71–80.

Crawshaw, M., & Montuschi, O. (2013). Participants' views of attending parenthood preparation workshops for those contemplating donor conception parenthood. *Journal of Reproductive and Infant Psychology, 31*(1), 58–71. https://doi.org/10.1080/02646838.2012.748886

Cushing, A. L. (2010). "I just want more information about who I am": The search experience of sperm-donor offspring, searching for information about their donors and genetic heritage. *Information Research, 15*(2). https://informationr.net/ir/15-2/paper428.html

Daniels, K. (2007). Donor gametes: Anonymous or identified? *Best Practice & Research Clinical Obstetrics and Gynaecology, 21*(1), 113–128. https://doi.org/10.1016/j.bpobgyn.2006.09.010

Daniels, K., & Douglass, A. (2008). Access to genetic information by donor offspring and donors: Medicine, policy and law in New Zealand. *Medicine and Law, 27*, 131–146.

Daniels, K. R., Kramer, W., & Perez-y-Perez, M. V. (2012). Semen donors who are open to contact with their offspring: Issues and implications for them and their families. *Reproductive BioMedicine Online, 25*, 670–677. https://doi.org/10.1016/j.rbmo.2012.09.009

Dempsey, D., Kelly, F., Horsfall, B., Hammarberg, K., Bourne, K., & Johnson, L. (2019). Applications to statutory donor registers in Victoria, Australia: Information sought and expectations of contact. *Reproductive Biomedicine & Society Online, 9*, 28–36.

Erikson, E. H. (1959). Identity and the life cycle: Selected papers.

Fertility Associates. (n.d.). *Information on the HART Act*. https://www.fertilityassociates.co.nz/media/1550/hart-info-for-patients-7.pdf

Freeman, T., Jadva, V., Kramer, W., & Golombok, S. (2009). Gamete donation: Parents' experiences of searching for their child's donor siblings and donor. *Human Reproduction, 24*(3), 505–516.

Freeman, T., Zadeh, S., Smith, V., & Golombok, S. (2016). Disclosure of sperm donation: A comparison between solo mother and two-parent families with identifiable donors. *Reproductive BioMedicine Online, 33*(5), 592–600. https://doi.org/10.1016/j.rbmo.2016.08.004

Golombok, S., Blake, L., Casey, P., Roman, G., & Jadva, V. (2013). Children born through reproductive donation: A longitudinal study of psychological adjustment. *The Journal of Child Psychology and Psychiatry, 54*(6), 653–660. https://doi.org/10.1111/jcpp.12015

Hargreaves, K., & Daniels, K. (2007). Parents dilemmas in sharing donor conception stories with their children. *Children and Society, 21*, 420–431. https://doi.org/10.1111/j.1099-0860.2006.00079.x

Hertz, R., Nelson, M. K., & Kramer, W. (2015). Sperm donors describe the experience of contact with their donor-conceived offspring. *Facts, Views & Vision in ObGyn, 7*(2), 91.

Isaksson, S., Sydsjo, G., Skoog Svanberg, A., & Lampic, C. (2014). Preferences and needs regarding future contact with donation offspring among identity-release gamete donors: Results from the Swedish Study on Gamete Donation. *Fertility and Sterility, 102*(4), 1160–1166. https://doi.org/10.1016/j.fertnstert.2014.06.038

Jadva, V., Badger, S., Morrissette, M., & Golombok, S. (2009a). 'Mom by choice, single by life's circumstance': Findings from a large scale survey of the experiences of single mothers by choice. *Human Fertility, 12*(4), 175–184. https://doi.org/10.3109/14647270903373867

Jadva, V., Freeman, T., Kramer, W., & Golombok, S. (2009b). The experiences of adolescents and adults conceived by sperm donation: Comparisons by age of disclosure and family type. *Human Reproduction, 24*(8), 1909–1919. https://doi.org/10.1093/humrep/dep110

Jadva, V., Freeman, T., Kramer, W., & Golombok, S. (2010). Sperm and oocyte donors' experiences of anonymous donation and subsequent contact with their donor offspring. *Human Reproduction, 26*(3), 638–645.

Kelly, F., & Dempsey, D. (2016a). The family law implications of early contact between sperm donors and their donor offspring. *Family Matters, 98*, 56–63.

Kelly, F., Dempsey, D., & Frew, C. (2019a). The donor-linking practices of Australian fertility clinics. *Journal of Law and Medicine, 27*(2), 355–368.

Kelly, F., Dempsey, D., Power, J., Bourne, K., Hammarberg, K., & Johnson, L. (2019b). From stranger to family or something in between: Donor linking in an era of retrospective access to anonymous sperm donor records in Victoria, Australia. *International Journal of Law, Policy and the Family, 33*(3), 277–297.

Kelly, F. J., & Dempsey, D. J. (2016b). Experiences and motives of Australian single mothers by choice who make early contact with their child's donor relatives. *Medical Law Review, 24*(4), 571–590.

MacDougall, K., Becker, G., Scheib, J. E., & Nachtigall, R. D. (2007). Strategies for disclosure: How parents approach telling their children that they conceived with donor gametes. *Fertility and Sterility, 87*, 524–533. https://doi.org/10.1016/j.fertnstert.2006.07.1514

Mahlstedt, P. P., LaBounty, K., & Kennedy, W. T. (2010). The views of adult offspring of sperm donation: Essential feedback for the development of ethical guidelines within the practice of assisted reproductive technology in the United States. *Fertility and Sterility, 93*(7), 2236–2246.

Michelle, C. (2006). Transgressive technologies? Strategies of discursive containment in the representation and regulation of assisted reproductive technologies in Aotearoa/New Zealand. *Women's Studies International Forum, 29*, 109–124. https://doi.org/10.1016/j.wsif.2006.03.009

Murray, C., & Golombok, S. (2005). Going it alone: Solo mothers and their infants conceived by donor insemination. *American Journal of Orthopsychiatry, 75*(2), 242–253. https://doi.org/10.1037/0002-9432.75.2.242

New Zealand Law Commission. (2005). *New issues in legal parenthood (No. 88)*. New Zealand Law Commission.

New Zealand Law Commission. (n.d.). Donor gamete conception. https://www.nzlii.org/nz/other/nzlc/pp/PP54/PP54-3_.html

Nordqvist, P. (2014). The drive for openness in donor conception: Disclosure and the trouble with real life. *International Journal of Law, Policy and the Family, 28*(3), 321–338.

Persaud, S., Freeman, T., Jadva, V., Slutsky, J., Kramer, W., Steele, M., ... Golombok, S. (2017). Adolescents conceived through donor insemination in mother-headed families: A qualitative study of motivations and experiences of contacting and meeting same-donor offspring. *Children & Society, 31*(1), 13–22. https://doi.org/10.1111/chso.12158

Ravelingien, A., Provoost, V., & Pennings, G. (2013). Donor-conceived children looking for their sperm-donor: What do they want to know? *Facts, Views and Visions in Obstetrics and Gynaecology, 5*(4), 257–264.

Ravelingien, A., Provoost, V., & Pennings, G. (2015). Open-identity sperm donation: How does offering donor-identifying information relate to donor-conceived offspring's wishes and needs? *Journal of Bioethical Inquiry, 12*(3), 503–509.

Readings, J., Blake, L., Casey, P., Jadva, V., & Golombok, S. (2011). Secrecy, disclosure and everything in-between: Decisions of parents of children conceived by donor insemination, egg donation and surrogacy. *Reproductive Biomedicine Online, 22*(5), 485–495.

Scheib, J., & Ruby, A. (2008). Contact among families who share the same sperm donor. *Fertility and Sterility, 90*(1), 33–43. https://doi.org/10.1016/j.fertnstert.2007.05.058

Scheib, J. E., Riordan, M., & Rubin, S. (2003). Choosing identity-release sperm donors: The parents' perspective 13–18 years later. *Human Reproduction, 18*(5), 1115–1127. https://doi.org/10.1093/humrep/deg227

Scheib, J. E., Ruby, A., & Benward, J. (2017). Who requests their sperm donor's identity? The first ten years of information releases to adults with open-identity donors. *Fertility and Sterility, 107*(2), 483–493.

Shaw, R. (2008). Rethinking reproductive gifts as body projects. *Sociology, 42*(1), 11–28.

Trail, R., & Goedeke, S. (2019). Going it alone: Stories of New Zealand women choosing single motherhood. *New Zealand Journal of Psychology (Online), 48*(2), 4–13.

Turner & Coyle. (2000). What does it mean to be a donor offspring? The identity experiences of adults conceived by donor insemination and the implications for counselling and therapy. *Human Reproduction, 15*(9), 2041–2051. https://doi.org/10.1093/humrep/15.9.2041

Webber, M. (n.d.). *Identity and Whakapapa: A curriculum for the gifted Maori child*. https://www.confer.co.nz/gnt/Thursday/webber.pdf

9

The Importance of a Genetic Link in Surrogacy Arrangements: Law, Public Opinion and Reconciling Conflict

Debra Wilson

Introduction

Clinic-assisted IVF surrogacy is the first form of assisted reproductive technology that has the ability to fragment genetics and gestation (known collectively as a biological link) from parenthood. In most other forms of assisted reproduction there is either a genetic link to a sole intended parent or to one of the two intended parents, or there is a gestational link. For the first time, this kind of surrogate pregnancy[1] allows for reproduction to take place where the intended parent(s) neither supply genetic material nor gestate the child resulting in no biological link to the intended parents; although it should be noted that emerging technologies might also allow for a similar situation.

This chapter will focus on the discussion of who the 'legal parents' of a child born through surrogacy are. As will be seen in the following paragraphs, this is a different question to what makes people 'family' or

D. Wilson (✉)
University of Canterbury, Christchurch, New Zealand
e-mail: debra.wilson@canterbury.ac.nz

'related', or what 'being a parent' means. These are important questions which cannot be discussed in appropriate detail in this chapter but have been the subject of detailed and interesting discussion elsewhere (e.g., see Bainham et al., 1999; Carsten, 2000; Probert, 2004).

From a legal perspective, clear identification of the 'legal parents' of a child is vital: it is through legal parents that a child gains rights of citizenship and human rights, and the legal parents are the ones who will exercise these rights and make decisions for the child until the child has reached such an age where they can do this for themselves. Identifying the 'legal parents' of a child in a surrogacy situation where multiple adults might be involved (these include the birth mother and her partner (if he or she exists and consents to the surrogacy), sperm and/or egg donors, and the intended parent(s) (if these are different from the genetic donors)), is a complicated but necessary step in any surrogacy arrangement. If the intended parent(s) is/are not also the legal parents then they must take steps to become so through some form of transfer of legal parentage (usually adoption, but in some countries alternatively through parenting or parentage orders). Without this step, the child will be left in an uncertain legal position and the intended parents will have no legal rights in relation to the child.

As suggested above, identifying the legal parents of the child can be complicated. Legal parentage can be defined through genetics (the male and female who provided the sperm and egg); through gestation (the woman who gives birth to the child is the mother, and her partner becomes the other parent if they have consented to the surrogacy) or through intention (the person(s) who intend to raise the child). Different countries have taken different approaches to this. In countries like France, Germany, Spain, Australia and England, a gestational model is applied. In countries like Portugal, Thailand and Taiwan, an intentional model is used. Other countries like South Africa and Greece use a model involving the judicial determination of parentage before the child is born.[2]

Traditionally, New Zealand has adopted a gestational model of parentage, but it has also gone further, requiring a biological (genetic) link for surrogacies which fall under the regulation of the Human Assisted Reproductive Technology Act 2004 (those surrogacies which involve the use of fertility clinics). It has recently been announced by the Advisory

Committee on Assisted Human Reproduction (ACART), however, that this will be removed. Consultation on this proposed amendment was limited, but ACART indicated that this would be a controversial change, with some respondents feeling strongly that a biological link requirement should be retained, and others (the majority) feeling strongly that it should be removed.

This chapter introduces the debate over the biological link policy in New Zealand and discusses the consultation on its removal. It then discusses some relevant results from a public perceptions survey on surrogacy carried out by the author, which provides some insight into the general public's thoughts and opinions on the importance of a genetic link between intended parents and the child. These results demonstrate that, as might be expected, the issue of genetics remains an important and divisive one in considerations of assisted reproduction policy.

Surrogacy Regulation in New Zealand

Surrogacy is specifically regulated in New Zealand through two provisions in the Human Assisted Reproductive Technology Act 2004 (HART Act):

Section 14: which states that surrogacy arrangements are not illegal but are not enforceable by or against any person; and which makes it an offence to give or receive, or agree to give or receive, valuable consideration for participation or arranging for participation, in a surrogacy arrangement. There are five stated exceptions to this.

Section 15: which makes it an offence to advertise in relation to participating in a surrogacy arrangement.

This is not, however, the extent of the regulation. Section 16 makes it an offence for anyone to perform an assisted reproductive procedure without the prior approval in writing of the ethics committee (ECART). To be clear: the regulation in the Act therefore applies to surrogacies where the pregnancy involves the use of a fertility clinic. It does not apply to 'private' surrogacy arrangements where the individuals take the relevant steps to achieve pregnancy in the privacy of their own homes. Under section 19, ECART may only grant approval in accordance with Guidelines provided by the Advisory Committee on Assisted Reproductive Technology (ACART). If there are no Guidelines, ECART cannot approve the application. Before Guidelines are issued, ACART is required under s39 to engage in public consultation, and to take this into account when considering whether to advise the Minister to make changes (in accordance with s35).

In 1997 ACART's predecessor, the National Ethics Committee on Assisted Human Reproduction (NECAHR) introduced guidelines relating to clinic-assisted surrogacy. These were subsequently adopted for use by ECART and remained the relevant guidelines until 2008, when ACART issued the 'Guidelines on Surrogacy Arrangements Involving Providers of Fertility Services'. These relevantly stated that:

2. When considering an application for a surrogacy arrangement involving a provider of fertility services:
 (a) ECART must determine that:
 (i) At least one of the intending parents will be a genetic parent of any resulting child.

In 2013 the Guidelines were revised following a complaint to the Human Rights Commission that the guidelines were discriminatory, and a new version was issued: the 'Guidelines on Surrogacy Arrangements Involving Assisted Reproductive Procedures'. These relevantly state that:

2. When considering an application for surrogacy involving an assisted reproductive procedure:
 (a) ECART must determine that:
 (i) Where there is one intending parent, he or she will be a genetic parent of any resulting child,

or

(ii) Where there are two intending parents, at least one will be a genetic parent of any resulting child;

...

In both versions, therefore, ACART has required that there be a genetic link between the child and at least one intended parent. This requirement is consistent with ACART's 'biological link policy', which requires "that a child born from an assisted reproductive procedure must have at least one biological link (either genetic or gestational) to an intending parent" (ACART, 2019, para 48).

Revisiting the Biological Link Policy: ACART's 2017 and 2018 Public Consultation

In September 2017 ACART released a consultation document, "Proposed Donation Guidelines: for family gamete donation, embryo donation, use of donated eggs with donated sperm and surrogacy" (ACART, 2017). This consultation sought feedback on whether to recommend changes to four Guidelines to the Minister, including the 2013 Surrogacy Guidelines.

ACART described several reasons for this consultation. First, a 2011 complaint to the Human Rights Commission had suggested that the then-current Surrogacy Guidelines were discriminatory on the basis of sex and sexual orientation. Second, although that complaint had been resolved through the publication of new surrogacy Guidelines, ACART became concerned that the new Guidelines were inconsistent with other existing Guidelines, requiring these to also be re-considered (ACART, 2017, para 20). ACART therefore initiated a review, in accordance with s35(1) HART Act (which requires ACART to regularly review its guidelines), to ensure that the other Guidelines were "consistent, where appropriate" with the new Guidelines, and at the same time to consider "the feasibility of having one set of guidelines to cover all four procedures" (ACART, 2017, para 21).

The consultation document sought feedback on 13 proposed changes to the guidelines, but acknowledged that the main issue was Proposal One:

> The guidelines should no longer require a gestational or genetic link between intending parents and a resulting child.

This was described as "the most significant policy shift" (ACART, 2017, p. piii) of the consultation, and as one of two "overarching matters" (ACART, 2017, para 48) (the second being the format of the guidelines). While acknowledging that there was a 'cultural dimension' to this proposed change, ACART commented that:

> The recognition of the centrality and importance of relatedness and connection to others expressed through values such as whānau, whakapapa and whanaungatanga,[3] is relevant to gamete and embryo donation. …
>
> ACART makes no assumptions, however, that knowledge of a person's genetic origins requires that the offspring born from gamete and embryo donation need be genetically related to their intended parents. Children born through assisted reproductive technology may still learn their whakapapa and whanaungatanga through gamete and embryo donation without the need for a biological connection. The 'donation' may also be the gift of carrying a child for another by means of surrogacy, which has parallels with the Māori customary practice of whāngai.[4] (ACART, 2017, para 25–26)

The document concluded that "there are potentially two factors more significant for the wellbeing of children born from third-party assistance than how an embryo is created or gestated" (ACART, 2017, para 25–26). These were described as "preparation before fertility treatment, when an individual or couple is looking at the implications for themselves and the child of having a child through gamete or embryo donation or using surrogacy" and "once a child is born, the way in which the family deals with the child's identity" (ACART, 2017, para 25–26).

The consultation document sought feedback on the following proposal:

> The guidelines should no longer require intending parents to have a genetic or gestational link to a resulting child instead the guidelines should require ECART to be satisfied that where intending parents will have neither a genetic nor a gestational link to a resulting child, the lack of such links is justified.
>
> Do you agree? Yes/No

> Do you believe there are any cultural implications associated with the proposed removal of the biological link policy? If so, please describe these implications.
>
> Please give reasons for your views.

There were 24 written responses to the consultation.[5] Twenty-two commented on the genetic link proposal. Of these, 14 were in favour of removing the requirement, 6 were opposed and 2 commented but did not state a firm opinion. Of those in favour, nine respondents specifically referred to discrimination or lack of access by certain person(s) to explain their answer. Five recognized the need for careful consideration of cultural matters. Three identified whakapapa and knowledge of genetic origins as important but noted that what is important is being able to trace or understand their genetic history, not necessarily being raised by someone with that genetic connection. Two commented that the nature of parenting was 'shifting' and that the current guidelines failed to recognize 'the complex reality of families today, some of which are not grounded in biological links'.

Of those opposed to removing the requirement, two commented that this was not in the best interests of the child, and that removal failed to show respect for the dignity of the child. One commented that merely being informed of genetic origins is not sufficient, what is required is 'lived knowledge'. One considered that this proposal was being 'driven by the fertility industry and consumer demand' and another felt that it was an 'overtly unlawful attempt at legislation'. One commented that should the child be born disabled, the intended parents would be more likely to accept a child where there was a genetic link. Another drew on adoption research to highlight that 'origins matter'.

In February 2019 ACART released a second consultation document (ACART, 2019), with the purpose of addressing issues that had emerged from the first consultation, and to re-consult on amendments that had been made following the consultation. This document confirmed ACART's intention "to progress the most significant policy shift, which is to rescind the mandatory biological link requirement" (ACART, 2019, p. piii).

While 58% (14 of 24) of respondents agreed with ACART's proposed removal of the genetic link in the 2017 consultation document, this is clearly a small sample size and respondents were, in general, interested parties (academics researching in the area, industry professionals and religious and advocacy groups) speaking in general terms rather than individual members of the public describing their personal views. It might not, therefore, represent the public opinion on the importance of the genetic link.

At the same time as the consultation was occurring, the author was designing a public perceptions survey with the aim of gaining insight into the public's knowledge, understanding and opinions of surrogacy in New Zealand. A specific question on the importance of the genetic link was therefore included in this survey to inform further debate.

The Public Perception: How Important Is the Genetic Link?

Methodology and Structure of Survey

In 2017–2018 the author carried out a paper-based survey of a representative sample of members of the New Zealand public in order to gain insight into public perceptions on surrogacy.[6] Participants were selected from the New Zealand General Electoral Roll, following a request to the Electoral Commission under s112 Electoral Act 1993 (which allows access to an electronic list of enrolled voters on the General Roll for members of specified state sector organizations for the purpose of human health or scientific research purposes). The supplied data separated voters into 10-year age bands. From this, a list was generated comprising approximately every 20th name in each band. Respondents with postal addresses outside of New Zealand were then removed from the list. Respondents were given the option of either filling out and returning a paper survey or accessing an electronic version. Paper-based responses received were manually entered into survey design tool Qualtrics for analysis alongside the online responses. Research ethics approval was given by the University of Canterbury, New Zealand Human Ethics Committee (ref. 2017/07).

Four thousand and forty-one surveys were posted out, with two hundred and eighteen returned marked 'not at this address' or 'deceased'. Five hundred and fifty-seven responses were received, which is a response rate of approximately 14%. While this is a low response rate, it should be noted that the electoral roll was obtained for this survey approximately two years after the most recent election (being the time when the public would have been required to confirm their postal addresses to the Electoral Commission). It was therefore recognized that people on the mailing list might have moved from their house and the new owners might simply have discarded the letter.

The survey asked 67 questions, separated into five parts: demographics; current understanding of surrogacy; past experiences with surrogacy; future plans in relation to surrogacy and thoughts on reform. Some of the questions in Part Five either asked specifically about a genetic link requirement or contained reference to this in their answers.

The Public Opinion on the Genetic Link Requirement

Question 63 asked, "*Should there be a requirement of a genetic link between at least one of the intending parents and the child born via the surrogacy arrangement?*" There were 500 responses to this question. One hundred and ninety-eight respondents (39.6%) answered 'yes'. Two hundred and thirty-six respondents (47.2%) answered 'no'. The remaining 66 (13.2%) answered 'unsure/prefer not to say'. The question then gave the option of making additional comments and 147 chose to do so.

The results and additional comments were filtered using demographic data collected in Part One of the survey. Filtering by the age of the respondent did not reveal any particular trends in answers. While there seemed to be a preference for a genetic link amongst the older age groups (43.5% of the 55–64 year group, and 52.9% of the 65–74 year group, answered 'yes') this was also the preference amongst the 25–34 age group (57.1% answered 'yes'). For the middle age ranges, 'no' was the most common response (55.6% of the 35–44 year group, and 53.8% of the 45–54 age group). Interestingly, these middle age groups are the groups most likely to utilize assisted reproduction.

Filtering by the education level of the respondents suggested that this had some impact on responses. Those who described their education level as being '*Less Than High School*' or '*High School*' appeared more in favour of a genetic link requirement (*Less Than High School*: 48.5% answered 'yes' and 42.4% answered 'no'; *High School Graduate*: 47.3% answered 'yes' and 39.6% answered 'no'), while those who had continued on to tertiary education appeared more in favour of no genetic link requirement (*Diploma*: 39.7% answered 'yes' and 42.7% answered 'no'; *Undergraduate Degree*: 32.7% answered 'yes' and 56.4% answered 'no').

The ethnicity of the respondents also appeared to have some impact on responses. Those identifying as '*Asian*', '*Maori*' or '*Pasifika*' were more likely to favour a genetic link requirement (*Asian*: 61.1% answered 'yes' and 16.7% answered 'no'; *Maori*: 39.1% answered 'no' and 30.4% answered 'yes'; *Pasifika*: 50% answered 'yes' and 33.3% answered 'no'). Those identifying as '*European*', '*NZ European*' or '*Other*' were more likely to favour no genetic link requirement (*European*: 53.8% answered 'no' and 33.3% answered 'yes'; *NZ European*: 48.6% answered 'no' and 39.1% answered 'yes'; *Other*: 61.1% answered 'no' and 33.3% answered 'yes'). It should be noted, however, that there was an anomaly in one age group. While *European* and *NZ European* respondents overall favoured no genetic link requirement, this was reversed in the 55–64 year age group (*Europeans aged 55–64*: 58.3% answered 'yes' and 25% answered 'no'; *NZ Europeans aged 55–64*: 45.2% answered 'yes' and 42.2% answered 'no'). While *Maori* respondents overall favoured a genetic link, this was also reversed in the 55–64 age group (*Maori aged 55–64*: 50% answered 'no' and 8.3% answered 'yes').

The gender of the respondents had no obvious impact on responses, with those identifying as either '*female*' or '*male*' both overall favouring no genetic link requirement (*Females*: 46.3% answered 'no' and 41.7% answered 'yes'; *Males*: 48.3% answered 'no' and 36.3% answered 'yes'). Further analysis, however, showed that this overall result was affected by the age of the respondents. Both male and female respondents who were under the age of 55 favoured 'no', but this reversed for the over 55 groups.

It should be noted that while a third gender option, '*other*', was available, no respondents to this question identified in this way.

The sexual orientation of the respondents had no obvious impact on responses. Those identifying as either '*asexual*', '*homosexual*' or '*heterosexual*' all overall favoured no genetic link requirement (*Asexual*: 53.8% answered 'no' and 38.5% answered 'yes'; *Homosexual*: 62.5% answered 'no' and 37.5% answered 'yes'; *Heterosexual*: 47.1% answered 'no' and 39.3% answered 'yes'). As with the analysis by gender, these overall results were affected by the age of respondents, with those under the age of 55 favouring 'no' and those over 55 favouring 'yes'. It should be noted that while a fourth sexual orientation option, '*other*', was available, no respondents to this question identified in this way.

Previous experience with fertility treatments appeared to have an impact on responses. Those respondents who described themselves as having previously '*entered into a surrogacy agreement*' or as having received '*donated eggs/sperm*' were overwhelmingly more likely to favour no genetic link (*Surrogacy*: 84.2% answered 'no' and 10.5% answered 'yes'; *Donated eggs/sperm*: 60% answered 'no' and 20% answered 'yes'), whereas those who had experienced '*other fertility treatment*' were overwhelmingly more likely to favour a genetic link (64% answered 'yes' and 13.8% answered 'no').

The question finally gave the respondents the option of adding additional comments, and 147 respondents did so. Some of those respondents who answered 'yes' considered that a genetic link would create a 'stronger connection' or a 'closer tie' or a 'biological commitment' to the child. One commented that they had an 'impression that step parents can be less committed' to the child. Another thought that if the child was born with a disability, intended parents without a genetic link would 'have no bonding, no genetic input, and could refute their obligations of responsibility for the care of the child'. Four respondents thought that a genetic link would be in the best interests of the child, suggesting that this link would provide 'an important human sense of belonging to be genetically connected somewhere' and others thought of it as 'strengthening the bond' or 'linking' the child to the intended parents. Three other respondents saw a genetic link as benefitting the family, helping with an understanding of 'genealogy' or reducing 'awkwardness in a family/whānau' or making 'traditional or cultural issues less troublesome'.

Six respondents saw the genetic link as important for broader societal reasons. Five thought that the presence of a genetic link would prevent surrogacy becoming a business, and one thought it would protect against the rise of designer babies. Many respondents simply commented that it would be 'preferable' (9 respondents) or 'ideal' (6 respondents) for there to be a genetic link but did not explain this further. Language of 'desirable', 'nicer', 'a good thing' and 'easier' also appeared in the responses.

Twenty-six respondents thought that if there was to be no genetic link, that adoption was more appropriate than surrogacy. One respondent noted, "no point [to surrogacy] otherwise—it's just the same as adoption if there is no genetic link". Others commented that there are already non-genetically related children needing homes, for example, "there are plenty of babies and children waiting for adoption, why add more?" and "adoption comes to mind—the child needing a home and a couple wishing for a child".

Finally, it should be noted that a significant number of those answering 'yes' qualified their answers in the comments by saying that a genetic link should be required only if it was 'medically' or 'socially' or 'physically' possible (11 respondents). Other respondents were more specific, identifying 'infertility' (9 respondents), 'sterility' (1 respondent), 'inability to conceive' (1 respondent) or 'having a genetic disorder' (4 respondents) as exceptions. Two respondents clarified that 'convenience' or 'philosophical' reasons should not be used to justify the lack of a genetic link.

The reasons of those respondents answering 'no' tended to fall into four categories. First, one group did not see the importance of a genetic link, for example, 'don't see why it matters' and 'the whole point of surrogacy is to relax the binds of genetics'. One even saw the lack as a positive: 'diversity is good'. A second group commented that genetics was not a prerequisite for good parenting, for example, "parenting is not dependent on genetics, just love" and "you don't have to be blood related to love a child and raise it". One respondent asked, "with IVF you can have egg and sperm donors that aren't genetically linked to either parent—is there a difference?" A third group compared surrogacy to adoption; for example, "non-genetic adoption works! Why not surrogacy?", "why artificially impede genuinely keen parents from being parents?" and "why should there be the question? Not required in adoption". A fourth group thought that genetics was a matter for the parties to the surrogacy arrangement to decide. One commented, "if people with no genetic connection can agree on an arrangement, it's not for me to object".

Other Discussions of Genetics in the Survey

While Question 63 was the only question to specifically ask about the importance of a genetic link, this did form part of discussions in other questions. Some of these will be discussed below.

Question 61: Legal Parentage

Question 61 asked, "*At present, the law states that the surrogate is the legal parent of a child born through surrogacy. Who should the legally recognized parents of the child born via a surrogacy arrangement be?*" This was an open question, with respondents given a text box in which to enter their answer. There were 498 responses to this question, with 103 writing 'unsure' or some equivalent.

Three hundred and ninety-five respondents identified one or more persons that they felt should be the legal parent(s). Of these, the most common answer was the 'intended parents' (207 respondents). The next most common response suggested that parentage depended on a genetic connection to the child (43 respondents). Only 21 thought the surrogate should be the legal parent. Other answers included some form of joint parentage (20 respondents), that parentage should be determined by the surrogacy agreement (18 respondents) or that it should be determined on a case-by-case basis (3 respondents). Twenty-four respondents took a slightly different approach, suggesting that the surrogate should be the legal parent until birth, and that this would then change (with 13 of these respondents specifically requiring adoption or some formal process of hand-over for this change to occur).

Throughout the answers, references to the relevance of a genetic link could be found. Of those who had nominated the intended parents as the legal parents, the answers of nine respondents appeared to assume that at least one of the intended parents was also the genetic parents. Three considered that if there was a genetic link to one intended parent, both intended parents should be the legal parents. Two others also considered the situation where only one intended parent was a genetic parent. Both saw it as problematic if the surrogate was the other intended parent, and thought this would result in the intended parents not being the legal parents. One added that if the other genetic parent was not the surrogate (i.e., a donor), both intended parents should be the legal parents. Three others thought that as long as the surrogate was not a genetic parent, the intended parents would be the legal parents (even if there was no genetic link with the intended parent(s)).

Of those who nominated the surrogate as the legal parent, none explained their reasoning, meaning that the importance of genetics could not be determined. Of those suggesting 'joint parentage' four took the approach that the legal parents should be the surrogate and any genetic contributors, one clarifying that this would include 'egg and sperm donors'.

Question 64: Cancelling the Arrangement

Question 64 asked, "*Should the law allow either the intended parents or the surrogate to cancel the arrangement once pregnancy has occurred? (i.e., can the surrogate decide to keep the child or can the intended parents decide not to take custody of the child?)*". There were 500 responses to this question. Two hundred and eighty-four respondents (56.8%) answered 'no', that cancellation should not be allowed. One hundred and twenty-three respondents (24.6%) answered 'yes' and were given the opportunity to specify circumstances in which cancellation could occur. Ninety-three answered 'unsure/prefer not to say'. Respondents were also given the opportunity to add additional comments, and 167 chose to do so.

Most of the answers used language of 'fairness' and 'certainty' as reasons to not allow either party to cancel the arrangement, or language of

9 The Importance of a Genetic Link in Surrogacy Arrangements… 217

the surrogate's 'autonomy' and 'best interests of the child' as reasons to allow cancellation. Some answers commented specifically that only one party should be able to cancel, with 13 thinking that only the intended parents could cancel and 51 thinking that only the surrogate could cancel. Four of the 51 specifically clarified that the surrogate should only be able to cancel if she had a genetic connection to the child. As an example, one thought, "it's slightly different if it is the surrogate's egg or not, i.e., if the surrogate is using their own egg I see it as more 'their' child than if the intended mother's egg. But it's a tricky area".

Questions 65 and 66: Changing Their Minds

Question 65 asked, '*If the surrogate mother changes her mind and wishes to keep the child, who should the court grant custody to?*' There were 508 responses to this question. Two hundred and sixty-three respondents (51.8%) thought that custody should go to the intended parents. Eighty respondents (15.7%) thought that custody should go to the surrogate, 65 respondents (12.8%) thought there should be joint custody, and 34 suggested some form of 'other' arrangement. Sixty-six respondents (13%) answered 'unsure/prefer not to say'.

Of the 34 that selected 'other', 24 explained this further. Nine respondents commented that genetics should determine custody. All of the respondents were then invited to explain their answer further. One hundred and seventy-two chose to do so. Of these, 44 respondents mentioned the genetic link as being relevant. Five respondents thought that the genetic link was the determining factor; for example, "depends on genetics" and "if the intended parents supplied the egg and sperm, the child is theirs". A further 17 thought that genetics was a relevant factor; for example, "I guess I feel 'ownership' of the child as such is partly determined by whose egg and sperm". Four respondents considered that a genetic link supported their belief that the intended parents should have custody; for example, "intended parents, particularly if have genetic link". Fifteen respondents who appeared to consider that the surrogate should have custody thought that a genetic link to an intended parent should result in a joint custody arrangement; for example, "surrogate mum plus parent(s) with genetic link". Finally, four respondents took a different approach, arguing that the lack of a genetic link to the surrogate should result in custody being given to the intended parents; for example, "no genetic connection, pure carrier, just like a caretaker of the child".

Question 66 asked, '*If the Intended Parents change their mind during the pregnancy or after the birth and do not wish to take the child, who should be responsible for: raising the child? child support?*' This was an open question, with respondents given a text box in which to enter their answer. Three hundred and sixty-five respondents provided comments. One hundred and fifty-two respondents thought that the intended parents should be responsible for the child. While most respondents did not provide reasons for their answers, 15 did give reasons. Twelve repeated the language of 'responsibility' to suggest that the intended parents should either raise the child or find adoptive parents. Two respondents mentioned the genetic link as relevant: "if these parents are genetic donors then they should have no right to cancellation or mind change" and "tough, it's their responsibility, their blood and their child". The next most common answer was that the surrogate should be responsible (66 respondents). Six respondents considered that the responsibility lay with the 'donor', 'genetic father', 'genetic parents', 'person with the most genetic attachment' or 'anyone with a biological link' to the child.

Commentary: How Important Is the Genetic Link?

The answers to the survey provide some interesting insights into the thoughts and opinions of members of the public in relation to a genetic link requirement. Some of these will be discussed below.

A Genetic Link Is Not Important...

As was discussed above, when asked directly whether people favoured a genetic link or not, the majority of respondents answered no, indicating that they did not see the genetic link between intended parent(s) and child to be necessary.

… Until It Is

The problem with asking a straight 'yes' or 'no' question is that people's opinions on a matter such as a genetic link appear to be more nuanced and complicated. While 39.6% of respondents answered 'yes' there should be a genetic link requirement, many then added comments which suggested that their preferred answer was 'yes, if possible' rather than 'yes, this should be mandatory'. This was seen through multiple comments which suggested that it would be 'preferable' but that recognized that there might be medical or social reasons why this would not occur. These answers became even more interesting when a scenario was introduced that involved conflict.

In Question 63, participants were asked whether either the intended parents or surrogate should be able to cancel the arrangement. It was described above how a significant number of people saw genetics as a way of answering this question. The analysis of the survey was taken one step further, however, with answers to this question being filtered in relation to their answers to Question 61. In other words, this question was considered: 'how did those people answering 'yes' or 'no' to a genetic link requirement respond to Question 63?' The results were unexpected.

Of those answering 'no' to a genetic link requirement, many commented on how difficult Question 63 was. One commented, "the wisdom of Solomon would be required to solve this situation—I'm not that wise". Some thought this was a 'minefield', contained 'fishhooks', was 'a can of worms' or a 'dilemma'. Others described it as a 'tough' or 'difficult" decision. In attempting to answer it, of the 87 of these respondents who added further comments, seven thought that genetics or biology was relevant to the answer. As examples; "Such a tough situation, I honestly don't know. I tend towards allowing the surrogate to keep the child if it was her egg, and not allowing the parents to. But so tough to determine" and "I believe that if one of the parents has contributed genetically then they are unable to cancel any agreement".

In comparison, some of those who answered 'yes' to a genetic link requirement focused more on the fact that this was an arrangement or contract between the parties, and that this should determine the answer. As examples, some wrote; "This is not buying a car" or "A child isn't a dress you can return if you don't like it" and "It is not like online shopping with a 30 day return period". Others focused on the child: "Once an agreement is made it should be upheld for the sake of the child", and "The child's best interests should drive the decision in this instance". Of the 63 respondents who added further comments, only two mentioned genetics: "The surrogate may have no genetic connection to the child so in that case, I believe should have no legal rights to the child" and "As long as there are good reasons, and depending on genetic relationships".

In Question 64, respondents were asked who the courts should grant custody to if the surrogate mother changes her mind. Of those who answered Question 61 by saying there should be no genetic link requirement, 81 gave further comments. Of these, 23 referred to genetics as a way of determining who should be granted custody. Where the surrogate was not a genetic parent, it was thought by these respondents that she should not be awarded custody; for example, "The surrogate is only carrying the child, and as someone provided DNA (egg/sperm) it would be questionable for the courts to grant custody", "No genetic connection, pure carrier, just like a caretaker of a child" and "Non egg donating surrogate is nothing [but] a human incubator".

A similar approach to this question was seen with those who answered Question 61 by saying there should be a genetic link. Seventy-one of these respondents provided further information, and 30 of these saw genetics as determinative (e.g., 'Dependent on genetics' and 'The surrogate is just the oven, not part of the child's DNA') or as an important factor in decision-making (e.g., 'Genetic makeup of child to play a big factor').

In Question 65 respondents were asked what should happen if the intended parents change their mind during the pregnancy. Of those who answered Question 61 by saying there should be no genetic link requirement, 99 provided further comments. Only three mentioned genetics in their answer, with the focus of the remainder on the feeling that the intended parents had responsibility to either raise the child or provide financial support, and acting in the best interests of the child.

The same could be seen with the respondents who had answered Question 61 by saying there should be a genetic link requirement. Of these, 85 provided further comments. Most answers focused on the intended parents having responsibility for the child (to either raise, financially support or arrange for adoption) due to the fact that they initiated the arrangement. Only four mentioned genetics, and this was again seen as a ground to recognize the responsibility of the intended parents to the child.

The answers to these results suggest that the issue of a genetic link is complicated. Of those who did not think it should be a requirement, a significant number used a genetic link as a way of addressing conflict. Of those who did think it should be a requirement, only a small number used it as a way of resolving conflict, preferring to rely on other factors as the primary way of addressing conflict.

The Presumption of a Traditional Family Construct

Although this cannot be measured, an impression was gained from reading the survey responses that participants had a default image of users of surrogacy as being a heterosexual married or de facto couple. Responses often referred to two parents, who were described as the 'intended father' and the 'intended mother'. This was despite the fact that a previous question in the survey had asked, "*Who should have access to surrogacy? (Please select all that you think should have access)*" and gave the options of: single males; single females; heterosexual married/civil union couples; heterosexual de facto couples; same sex female married/civil union couples; same sex female de facto couples; same sex male married/civil union couples; same sex male de facto couples; unsure/prefer not to say.

Some respondents commented that they found it difficult to imagine a situation where both intended parents were infertile. This suggests that while most respondents appeared to be familiar with the concept of surrogacy as a form of assisted reproduction, they were not necessarily familiar with all of the potential reasons for considering surrogacy. These might include medical reasons other than fertility; for example, having a condition affecting the ability to gestate and give birth, or having a genetic condition that they do not wish to pass on to a child. They might also include social reasons; for example, being single, in a homosexual relationship, or asexual, or situations where a person is capable of being a genetic donor or gestating a child, but where this would not fit with their self-image (e.g., a transgender or intersex person, or a female in a lesbian relationship who identifies with a more masculine role).[7] This confusion by some respondents suggests that while the general public appears to have some understanding of surrogacy, more can be done to educate as to the reasons why people consider this as an option for forming their families.

In Question 61 in particular, there seemed to be a commonly held assumption that at least one intended parent would be a genetic parent as well. Some respondents showed this explicitly, through comments that the child is 'their child', referring to the intended parents, and several used the term 'incubator' to refer to the surrogate, suggesting they did not think there would be a genetic link there. Other answers showed this more implicitly, with comments that suggested they saw the genetic parents and the intended parents as the same people; for example, 'Genetic intended parents should be the legal parents', or using the terms 'genetic parents' and 'intended parents' interchangeably.

These points suggest that answers might differ depending on the understanding of potential participants in a surrogacy arrangement held by a respondent. While the ability of surveys to both educate and seek feedback is limited, some participants mentioned that the questions asked them about issues they had not thought about, or that they had now changed their opinion on earlier questions having considered the issues raised in the later question. One commented that they wanted to go back and change earlier answers but thought that perhaps the survey was designed to deliberately give them knowledge as it went on, and so did not want to affect the results by changing position. It would have been useful to have planned to carry out some focus group interviews to explore the opinions of respondents in greater depth.

Conclusion

This chapter has attempted to contribute to an ongoing debate in the international literature on the importance of a biological link requirement in assisted reproduction. It has focused on surrogacy arrangements in New Zealand, where for the first time an assisted reproductive technology allows for parenthood to be fragmented into intention to raise and biological contributors, and considered the responses from a consultation by ACART on retaining a biological link requirement, and a public perceptions survey on surrogacy carried out by the author.

Both the ACART consultation and Question 63 of the public perceptions survey asked whether a biological or genetic link should be required in family formation, and found that the majority of respondents answered 'no'. While this provides important information, it does not give a complete picture of the importance that the public places on biological or genetic links. While some respondents to the survey did not see a genetic link as important in the decision to form a family, they nevertheless saw it as an important consideration should conflict or disagreement arise after the initial decision. When asked to consider scenarios where the surrogate or intended parents changed their minds (the surrogate wants to keep the child or the intended parents do not want to take custody of the child), the language of genetics featured significantly in the answers as to how to determine custody or parental responsibility. This was especially noticeable amongst those respondents who had initially said under Question 63 that a genetic link was not important for family formation, although most of the participants in this group did not elaborate on their thinking in relation to the two questions.

While the responses to these questions might initially seem inconsistent, there might be a logic behind it. Given a situation where the surrogate changes her mind and wants to keep the child, and the issue is before the court, a decision will have to be made in relation to custody. In trying to imagine what the court might take into account, respondents identified three approaches: 'best interests of the child', which is clearly hard to determine and might be resolved by considering the financial situation of the parties; 'enforcing the contract/agreement', which can appear like the child is being treated as an object being bought; or 'genetic link'. Focusing on the genetic link as a way of determining custody/responsibilities in situations where there is a conflict has the obvious advantage of basing the decision on a scientifically verifiable measure. In other words, a person is either genetically related to the child or not, whereas asking who will be the better parent will be incredibly difficult to determine on the facts and will rely on subjective views of the judge.

The importance of a genetic link in family formation and in ongoing family relationships is not a simple question, nor is it the only question. It should not be considered in isolation from the question of how custody/responsibility issues might be resolved in cases of conflict. Overall, therefore, while the ACART consultation should be applauded for recognizing that a biological link requirement limits (probably unnecessarily and discriminatorily) access to assisted reproductive technologies, the responses to the public perceptions survey act as a reminder that the role of genetics and biology is a complicated issue that has many layers to it and is one that might require ongoing consideration as the use of surrogacy and other emerging assisted reproductive technologies continues to grow.[8]

Notes

1. Both clinic-assisted IVF surrogacy and traditional surrogacy, where a woman provides the ovarian egg and gestates the foetus, are common in New Zealand. In this chapter, the term 'surrogacy' is used as shorthand for gestational clinic-assisted surrogacy. See Chap. 10 in this volume for a discussion of traditional surrogacy.
2. For an international survey on surrogacy laws worldwide, see Scherpe et al. (2019).
3. 'Whānau', 'whakapapa' and 'whanaungatanga' are Maori words in common usage in New Zealand. 'Whānau' can be understood as an 'extended family, or community of related families who live together in the same area', 'whakapapa' can be understood as a 'genealogy or line of descent from ancestors' and 'whanaunatanga' can be understood as 'kinship or a close connection between people'.
4. 'Whāngai' is also a Maori word in common usage in New Zealand. It can be understood as a 'cultural' or 'informal' adoption of a child within the extended family.
5. These submissions are available at https://acart.health.govt.nz/consultations/past-consultations.
6. Approval for this project was given by the University of Canterbury Human Ethics Committee (reference number HEC 2017/07).

7. See, for example, the submissions to the *Proposed Amendments to Guidelines on Surrogacy Arrangements involving Providers of Fertility Services and Guidelines on Donation of Eggs or Sperm between Certain Family Members* (2012), available at https://acart.health.govt.nz/consultations/past-consultations.
8. It should be noted that shortly after writing this chapter, the New Zealand Law Commission announced that the Minister Responsible for the Law Commission, Hon Andrew Little, has referred 'a review of surrogacy laws' to the Commission as part of their 2020/21 work agenda. See Te Ake Matua o te Ture (2020).

References

ACART. (2017). *Donation guidelines: Proposed guidelines for family gamete donation, embryo donation, use of donated eggs with donated sperm and surrogacy: consultation document.* Advisory Committee on Assisted Reproductive Technology. https://acart.health.govt.nz/system/files/documents/pages/review-donation-guidelines-consultation.pdf

ACART. (2019). *Second round of consultation on the proposed donation and surrogacy guidelines: Further changes since ACART's 2017 consultation.* Advisory Committee on Assisted Reproductive Technology. https://www.health.govt.nz/publication/second-round-consultation-proposed-donation-and-surrogacy-guidelines-further-changes-acarts-2017

Bainham, A., Sclater, S. D., & Richards, M. (Eds.). (1999). *What is a parent? A socio-legal analysis.* Hart Publishing.

Carsten, J. (2000). *Cultures of relatedness: New approaches to the study of kinship.* Cambridge University Press.

Probert, R. (2004). Families, assisted reproduction and the law. *Child and Family Law Quarterly, 16,* 273–288.

Scherpe, J. M., Fenton-Glynn, C., & Kann, T. (Eds.). (2019). *Eastern and Western perspectives on surrogacy.* Intersentia.

Te Ake Matua o te Ture. (2020, August 4). *The Law Commission's 2020/21 work program.* https://www.lawcom.govt.nz/news/law-commission%E2%80%99s-202021-work-programme

Part III

Relationships

10

Surrogacy and the Informal Rulebook for Making Kin Through Assisted Reproduction in Aotearoa New Zealand

Hannah Gibson

Introduction

Of all the assisted reproduction options, surrogacy is the least practised and the most elusive to those who consider it as a pathway to help create a family. Gestational surrogacy, where the surrogate is implanted with a non-genetically related embryo in a fertility clinic, and traditional surrogacy, where the surrogate is typically inseminated with the sperm of the intended father at home via syringe (and is genetically related to any baby she gestates), are permitted in Aotearoa New Zealand (hereafter, New Zealand), on an altruistic basis (Shaw, 2008). While those practising gestational surrogacy have a degree of medical and ethical guidance before their search for a surrogate and during the ensuing periods, those undertaking traditional surrogacy mostly do so outside of any formal guidance. For many people this is both liberating and potentially risky as they take on the technical intricacies and risks of home inseminations alongside

H. Gibson (✉)
Te Herenga Waka–Victoria University of Wellington, Wellington, New Zealand

having to find matches with suitable parents or surrogates. People participating in traditional surrogacies are also less likely to seek private counselling (which is compulsory for gestational surrogacy) before they begin their fertility journeys because it is seen as an extra cost or unnecessary psychological requirement. Although there are differences between these two types of surrogacies, both occupy a nebulous space wherein surrogates and intended parents (IPs)[1] co-constitute vulnerability. Surrogates and IPs experience uncertainty around how to find their "match" (i.e., locate a suitable surrogate or intended parent), the intricacies involved in developing relationships, and cultivating trust with each other when they agree to begin their reproductive journey. The complexities of surrogacy are manifold as they try to make kin in ways and spaces that have been typically deemed 'non-traditional'.

Riggs and Due (2013, p. 1) suggest that in Euro-American kinship, 'reproduction via heterosex remains the most valued form of reproduction', and people who reproduce outside of normative spaces due to social or medical infertility experience 'reproductive vulnerability', regardless of sexuality. On the surface, it appears that those who practise surrogacy in New Zealand have no reference point to guide them. However, in their search for support, guidance, and a potential match, many people turn to a virtual community that comes together on the internet, where different support forums become a central part of their journey. With no social guide on 'how to find a surrogate' in a country where couples are prohibited from publicly advertising for someone to carry their baby, people manage their reproductive vulnerability by developing ways to navigate the uncertainties they face in a legally ambiguous landscape.

Research Context

In New Zealand, it is illegal to advertise for a surrogate, and laws that govern the practice (such as what counts as compensation for expenses) are vague (Shaw, 2008, 2020), leaving both IPs and surrogates vulnerable to a degree. To proceed with gestational surrogacy, people must meet certain medical criteria and obtain approval from the Ethics Committee for Assisted Reproductive Technologies (ECART), a ministerial committee that approves, declines, and monitors each surrogacy application. In comparison, traditional surrogacy does not fall under the remit of the fertility clinic or ECART since it is not considered an assisted reproductive technology. Commercial surrogacy is prohibited, and at the time of writing this chapter, regardless of which adults are the genetic parents, the surrogate mother and her partner (if she has one) are the legal parents of the child (Shaw, 2020). The intended parent/s must file an adoption order to transfer the parental rights to them, which can take approximately six months.

In other jurisdictions such as the US, surrogacy agencies mediate these arrangements and control the matching process (Berend, 2016). More recently, Berend and Guerzoni (2019) indicate that online matching and in-house surrogates within fertility clinics have de-centred the monopoly that agencies once had over the process. In New Zealand, which has no history of agency mediation between actors, IPs using both traditional and gestational surrogacy arrangements rely on friends or family offering to carry a baby for them. The majority of people turn to Making Babies through Surrogacy (MBS), the central closed online forum where the majority of surrogates and IPs meet and match. Although some of my participants met through other platforms (social media, campaigns, real-life and other online groups), this chapter focuses on MBS because of its centrality in their lives and the influence it has on how people think about relationships and surrogacy more generally. The forum is a private space where people who are interested in doing surrogacy, and donor insemination to a lesser degree, can get guidance and support from those who have already been through it.

This chapter is based on ethnographic fieldwork with the New Zealand surrogacy community between 2016 and 2019. The moderators of the private online surrogacy group were happy to post my invitation to take part in the research project, which led to finding key participants, including 20 surrogates and their families, and 20 IPs throughout New Zealand. As I did not conduct online observation on MBS, my knowledge of how the forum works was gleaned from conversations with my participants. I conducted semi-structured interviews, attended medical appointments and procedures, baby showers, scans, and was lucky to be invited into many of my participants' homes and got to know their families. Everyone in the study, except two surrogates, identify as Pākehā (New Zealand non-Māori, usually of European descent), more than half are traditional surrogates (with two having been both traditional and gestational surrogates), and of the IPs, one was a gay couple, and one intended parent was single. Five sets of IPs went overseas for international surrogacy. With traditional surrogacy practised outside of any regulation or surveillance in New Zealand, it is almost impossible to know how many of these arrangements exist. Anecdotal evidence from several of my participants suggests that approximately 50% of surrogates in New Zealand are traditional. For this research I obtained research ethics approval from the Victoria University of Wellington Human Ethics Committee on 26 May 2016 (Reference number 22968). Pseudonyms have been used in the following discussion to protect the privacy of study participants and the forum's confidentiality.

Guidelines and Rules for Navigating Surrogacy in New Zealand

Over the course of my research, I saw patterns where surrogates and IPs modelled their search on another process they were familiar with: online dating. Rituals and language associated with the dating world are employed by members of the surrogacy community to give them a sense of control in response to the precarious aspects of assisted reproduction. From their search online for a partner to agreeing to go on a first 'date', relationships are carefully crafted with an unspoken assumption that there is a 'recipe for success'. The 'right' behaviours that are promoted in the community include who can make first contact and making sure to only date one potential match at a time. The 'wrong' behaviours can result in the removal of members from the forum. In this chapter, I show how both desirable and undesirable behaviours are judged in a variety of ways and how these shape informal rules around traditional and gestational surrogacy in New Zealand. The idea that following the rules will lead to a successful surrogacy journey helps people mobilise a set of interaction guidelines and a particular form of legality, what I call *shadow-legalities*. I propose that these codes of behaviour conceptually operate to offer the differently positioned participants a sense of security and belief in the 'system'.

Anthropologists who research 'shadow spaces' have largely looked at informal economies and, in contrast with alarmist ideas of illicit networks that exist primarily underground and disconnected from legitimate economies, they suggest that in today's globalised political-economy, the formal/informal, licit/illicit, legitimate/criminal often merge (Ferguson, 2006; Galemba, 2008; Sampson, 2003). Nordstrom (2000, p. 36) uses the term 'extra-state' to draw attention to how these economies exist simultaneously yet operate as distinctive politico-economic systems and authoritative mechanisms. Importantly, shadow networks are not necessarily illicit, but a space where culture is created and has social principles that govern conduct and exchange (Nordstrom, 2000, p. 37).

Rather than focus on the traditional interpretation of shadow spaces as a globalised movement of goods and services across borders, *shadow-legalities* describe how people navigate and respond to the ambiguity found in local institutional and intimate landscapes of kin practices in the context of New Zealand surrogacy. As they create kin in the shadow of heteronormative reproduction (even if certain practices or beliefs work to reinforce normative kinship), actors must find ways to establish their own informal rules and codes of conduct. As a set of salient and explicit rituals and narratives, *shadow-legalities* and the interaction guidelines that operate alongside them work as a set of parameters to frame the practices my research participants use to construct successful surrogacy journeys. These guidelines of engagement are established and reinforced by senior members of the surrogacy community as informal rules, providing new members with a sense of agency and hope. Ultimately, by creating their own quasi-regulatory and surveillance system by abiding by communally agreed upon rules, actors have a malleable road map of sorts to manage their practical and intimate relationships. This means that people interpret advice and rules of engagement to suit their own goals, and while rules are not strictly enforced, actions are monitored and judged.

At a deeper level, the concept of *shadow-legalities* is useful to understand how everyday rituals of kin-making in the surrogacy community are inspired by and embedded in official laws and legal processes. For example, on the surface, surrogacy is a mode of building and maintaining kin functions outside of the traditional symbolic logic of heteronormative reproduction and the law (Goodfellow, 2015). However, underpinning peoples' rituals and narratives within the surrogacy context is often beliefs that seek to work on and around the law to validate their kin-making. Although *shadow-legalities* do not have the force of the law, they are used to cultivate trust between parties rather than erase vulnerabilities inherent in surrogacy itself. They help people to navigate both the surrogacy community and satisfy the requirements of the formal legal system they must engage at various stages throughout the reproductive journey. For example, when parties demonstrate that they know one another properly and have taken the time to discuss the risks around surrogacy arrangements, as well cultivating a good relationship, this is valuable for both the ECART application process and mandatory engagement with the Ministry of Social Development in the lead up to the adoption process of the baby by the IPs. This is not a foolproof framework, of course, but the rules have been created in response to institutional and ethical requirements. Instead of functioning as two separate spaces, as the public/private or legal/illegal, the relationship between state regulatory guidelines and the informal rules and network formed by members of the surrogacy community can be blurry and overlap at times. In the following sections, I unpack the beliefs that underpin unspoken protocols or expectations around these *shadow-legalities*, including how the concept of romance functions as a way for people to connect in a non-legal way as they get to know one another.[2] I use four 'informal rules' outlined by the surrogacy community to thematically organise the data I present. These are rules that the members of the community feel that they must abide by, in order to belong to the community and to be successful in their surrogacy journeys.

During fieldwork, I saw a copy of the initial rules people are required to agree to when they apply for the forum. To respect confidentiality of the forum rules I have chosen to use the phrasing and wording that my participants use, as it is their interpretation of the rules and how it influences their journeys that I focus on here. The first rule is: following the rules helps your personal surrogacy journey; the second rule requires a leap of faith if you want to be successful; the third rule prescribes 'socially acceptable' norms around exclusivity when on a surrogacy journey; and finally, the fourth rule offers guidance on how to mitigate risks inherent in some surrogacy arrangements. The combination of these rules helps shape the guidelines and *shadow-legalities* that kin-making requires for surrogacy arrangements in New Zealand.

Study Findings

Rule One: You Will Be More Successful If You Follow the Community's 'informal rules'

I had the urge to impress Heather, who runs the MBS forum with help from other senior members. This was important, not only because Heather is well respected in the community and people seek her advice, but also because she was the proverbial anthropological gatekeeper. If the interview went wrong, it would surely impact on whether other people wanted to talk to me. Fortunately, when I arrived at her house, she opened the door, with a welcoming smile that put me at ease within moments. I quickly felt like an old friend sitting down at her kitchen table rather than a researcher collecting data. I felt the door being opened to me and being welcomed into the folds of the surrogacy community—a place where I would not only learn a lot but could potentially give back through my research.

Heather was forthright when we began discussing the dynamics of MBS. No 'voyeurs just reading and not posting', 'journalists after a juicy story', nor 'people with bad intentions' could join or stay if discovered, she said. This is primarily because the group is a haven for its members where they will often share details of their lives that even their extended family or friends do not know. For IPs, because they cannot have children via normative conception, they are reproductively vulnerable. At the same time, even though they get to know others who understand their experiences, it is still high stakes for IPs especially, because although there are other domestic forums for New Zealand IPs to find surrogates—MBS is the busiest and thus most promising.

While participants understood that there are informal rules on MBS, it became apparent to me early on in my fieldwork that the language they used when discussing their experiences was reminiscent of my own experiences of searching for love online in the past. In the following sections, I examine the interplay between the emotions people experience as part of the surrogacy arrangement process and the importance of rule following as key to surrogacy journey. In doing so, I highlight how certain processes and informal rules on familiar pathways that legally recognise relationships, like marriage, enable the formation of *shadow-legalities*. These paradoxically work to make people within the surrogacy community feel more in control, which is underpinned by the assumption that if they follow the rules of what other, already matched parties did then they too might be successful. These informal rules that enable local *shadow-legalities* are not so much created as they are transposed from established approaches to love and procreation.

Online environments are spaces where uncertainty is heightened, be it matching with a potential romantic partner (Gibbs et al., 2011) or locating a surrogate or IPs (May & Tenzek, 2016). Both scenarios also have high stakes if they have the goal of creating kin, in whatever way it may manifest. In my research I discovered that my participants, who were all seeking platonic relationships with strangers that would end in the surrogate gestating the baby for the IP/s, all found the terminology that is traditionally used in Euro-American dating and romantic terminology useful as a framework to approach and describe their experiences and feelings.

With more people wanting babies than those willing to be surrogates in New Zealand, I was curious about the specific process of finding a match on the forum.

> Madison: On the site there's rules there. It's got to be the surrogate approaching the IPs. The IPs can't just ask surrogates, "Will you have my baby?" That would be awkward.
> Hannah: What is the process of finding IPs?
> Madison: It's easier for the surrogates because when we join, we can look at all the discussions everyone has had before and look to see whose personality we connect with, or who we want to know better. It's very matched like

a dating thing. You see a couple you might like the look of. It is a relationship. You might tentatively say, "Would you like to meet for coffee or something? Have a chat?" No obligation.

Madison is a traditional surrogate who gestated siblings for her IPs. Her narrative reveals the power dynamics at play for new members, and the way rules around initial contact off the forum impact surrogates and IPs differently. '[For IPs], it's not just good enough to log on, put up your advert, sit back and wait for the babies to roll in because it ain't gonna happen', Carl, an IP via traditional surrogacy, explained. He refers to the unspoken rule: to be active. Tessa, an IP via gestational surrogacy, similarly stated, 'you have to actually make an effort and engage in conversations with other people and not come on and assume you'll automatically have someone go, "ok, I'll have your baby for you". [You have to] build up the rapport and friendships'. If a 'newbie' (a person who is new to the online forum) joins and claims to have read the rules, but immediately posts an advertisement and messages surrogates privately, then Heather quickly removes both, '[and I ask] them to read the rules to see what they did wrong. Usually they apologise and its fine. Strike two they are out though—if they continue to try and message surrogates'. Enforcement of the rules is strictly adhered to for the purpose of maintaining a safe space for everyone and to protect the surrogates from being bombarded by offers. Offensive comments are also prohibited, and promptly 'deleted and dealt with via private message—[and the] member will get deleted if it happens again' (Heather). Heather was also quick to point out that, 'I would never tell someone not to go ahead with someone—even if it's a huge red flag to me—I give them the information and they can do with it what they will'.

The enforcement of rules via the removal of messages and warnings demonstrates the importance of a continuity within the forum culture that people follow and reflects the risks involved. Berend (2016, p. 22) similarly found in the US context that respect is often gained by those who 'do their homework' and try to fit in. IPs must therefore be patient, become invested in others' journeys by commenting on their posts, and wait for at least one month before they can put up their own advert asking interested surrogates to contact them. IPs are essentially 'vetted' by the surrogates from afar. They create an idea of what an IP is like as a person by the way they interact with others, signalling the weight such actions have on reproductive opportunities. At the same time, much like Berend's study, senior forum members try to steer people in the right direction. Beth, a gestational surrogate, became a go-to member of the community for people with worries or questions. For her, it is vital that people do not rush into anything, particularly when it comes to traditional surrogacy, because this could have negative consequences for both the relationships themselves and the wider surrogacy community. The rules must be followed, to stay in the community.

The power dynamic and initial rules can be daunting for some IPs, even within the parameters of the rules. Sonia, who had a hysterectomy as a result of cancer treatment in her early thirties, but was able to keep her ovaries, found that even if she sparked up a friendship with someone on the forum, she felt too awkward to ever bring up the topic of surrogacy in case she was seen to be using someone just for their womb. 'In my reading of the law… surrogacy is something that you can't put pressure on anybody to do, so me asking anybody or even bringing it up is an indication of pressure, I think'. Although this central tenet allows people to have something concrete to do in response to uncertainty they might feel, Sonia's experience indicates that it creates another sense of anxiety and influences behaviour, particularly the IPs. The pressure can take its toll, as Christina (an IP via traditional surrogacy) found, 'I would spend at least an hour a day of my working day, researching and creating relationships and working on my profile, encouraging others and being there for others, building relationships. All I thought about was surrogacy and the fear that no one would pick me'. This profile maintenance is reminiscent of dating sites and how people must craft an ideal and attractive self. In the end, Christina left the site not long after, feeling discouraged because she received no interest or offers from a traditional surrogate. For Lola's second set of IPs, beyond being active on the forum, they were lucky that others from the forum already knew them in real life:

> With [my second IPs], she's super active on the forum, there are people that have met them, who knew that they were decent people… that was why I narrowed it down to them, but definitely being active on the forum

was really important because you get more of a picture of who they are and the fact they're not just, 'Hey, I wanna baby, come and find me!' kind of thing.

At the same time, some surrogates are aware of the unbalanced power dynamic. Lola was uncomfortable and worried that IPs feel the pressure to say 'yes' to any offer they get, whether there were red flags or not. As Heather said, 'For surrogates it's easy, there will always be another IP around the corner. That's why I [try] not to gloss over or make myself [look like someone I'm not], cause there's no point, I'd be wasting both of our time'. At the same time, surrogates also have to negotiate unknowns, even if they have done it before, because each relationship is new. 'It may seem that IPs would be [more vulnerable] initially... IPs can be lied to, strung along, ripped off, but then surrogates can be lied to, strung along and ripped off just as much'. Surrogates shared with me their fears that their IPs may expect them to get pregnant after the first implantation of the embryo/home insemination and wondered if they will have support during the pregnancy and/or a difficult birth. The many unknowns between IPs and surrogates produce co-dependencies around these informal rules to which members adhere to remain a part of MBS and have ideally predictable (re)productive surrogacy arrangements.

Rule Two: You Have to Take A Leap of Faith

> I really liked them, so that evening [after the event], cause I never take long to do things, I emailed Celeste to say 'oh, you know, I'm kind of in the market for an IP, and I really liked you and Adam, and do you wanna get to know each other... and see if this is something that could work?' (Lola)

Alongside active engagement and ensuring not to overwhelm surrogates, another popular piece of advice is to not 'jump into' surrogacy arrangements. According to Heather, if you go in with 'guns blazing', then there is a huge possibility it will go 'pear-shaped'. Similarly, according to Beth, a gestational surrogate:

Yeah, when people have rung me from the forum, and they've met someone and they're gonna start, and I'm like "oh, my god this is just a disaster waiting to happen. It's gonna ruin it for the other people as well". Jumping into bed before you've said "hello".

Despite this, there was a tension between advice not to rush things, and the reality that many surrogates choose their matches based on intuition and oftentimes quite quickly, as seen in the opening narrative. When Lola met the couple who would quickly become her first IPs, she did not see any point in waiting because she felt they would be a great fit. Heather echoed the sentiment; 'It's a feeling', referring to her instinctual decision to be a traditional surrogate twice. Madison likened the first meeting she had with a gay couple she would go on to carry two children for via traditional surrogacy as 'kinda like love at first sight', adding:

[My IPs] are so easy to talk to and we just seemed to have the same ideas about stuff. We met for dinner and coffee and stuff a few times, but I pretty much offered straight away… on our first date. It was easy… I think I was probably quite lucky that they weren't weird.

According to Berend (2016, p. 66), US surrogates discuss the importance of getting to know prospective IPs before making a decision, yet at some level they privilege 'a leap of faith', relying on their gut instincts. Berend (2016) describes online conversations where surrogates use the phrase 'it was meant to be'. Similarly, within the Israeli context, Teman (2010) discusses how people rely on intuitive connections as a basis for choosing one another. In New Zealand, I propose that 'taking a leap of faith' functions alongside pragmatism as a rule because it helps members to navigate notions of choice and calculation in the matching process. People are expected to engage in both universal behaviours, like IPs not being allowed to contact surrogates, while relying on their own intuition to ultimately choose a match. Whether it turns out well or not is another factor, as I explore in the discussion section below.

An example of combining intuition and pragmatism is evident in Tracey's situation. Tracey, a 28-year-old single mother of 3 girls, offered to be a traditional surrogate to Louise and Mike, who were in their early fifties. They had tried to carry a baby to term for many years, followed by using donor eggs, without any success. I (Hannah) was lucky enough that when I flew into the region where Tracey lived, Louise and Mike were there for a scan. I met the latter two first, on the evening at a local coffee shop before I was due to attend a local morning tea for those connected to surrogacy and meet several potential participants. When Louise and Mike arrived and sat down, they looked tired but had smiles from ear to ear, still in shock all these months later that their surrogate chose them. Almost serendipitously, Louise was very close to withdrawing from the forum due to no luck in finding a match, until she got a message from Tracey. When I interviewed Tracey the next day and asked her why she chose Mike and Louise as her IPs, she was candid:

> Aww, Louise's posts, because I've read through quite a few different ones and initially I thought they were too old, they were outside of my age spectrum, but of all the comments I read, hers was the most upbeat, and she commented lots on other peoples' posts, so she was very supportive of other people and very involved and I was like well, her responses sound like something I would say, so I re-evaluated to myself. I was like well, how much does age matter, and how much does it matter these days when people are living for longer anyway, with better health?

Despite initially thinking Louise and Mike were not a suitable match due to their age, Tracey changed her mind and offered to be a surrogate as a direct result of Louise's engagement on the forum.

Both Berend's (2012, 2016, p. 65) research on a surrogacy forum in the US and Teman's (2010) research with gestational surrogates in Israel identify similar ways surrogates compare matchmaking with IPs to dating, where the language of love echoes a common ground everyone is familiar with. As Berend puts it, 'This language, learned and internalized, creates both a cultural conceptualisation of surrogacy and a ground for action' (2012, p. 914). Reiterating this point in a later discussion, Berend goes on to say that romantic narratives used by many surrogates and IPs 'situate [their] interaction[s] in the intimate sphere, where baby making belongs' (2016, p. 67). Given that my participants endure an uncertain and physically and emotionally intense journey together that eventuates in kinship being created, it was unsurprising to me that many describe their relationship with one another with intimate, romance terminology.

Helena (1994) also found a similar use of romance terminology in narratives on her research in the US context, the difference being that at the time of her research surrogacy agencies did the matching. Berend's (2012, 2016) more recent work better compares with the New Zealand online surrogacy forum context because the public online group where she conducted her ethnographic research demonstrates a move from (or expansion of) agency focused organisation to a space and context where surrogates work collectively to talk about the cultural and emotional significance and implications of their roles. In particular, the surrogacy forum that Berend bases her research on has similar features to MBS, such as older members remaining on the forum to advise and guide newer surrogates. The informal rules provided by one societal framework—online dating—allow constitution of another set of informal rules in spaces where kin-making via surrogacy is undertaken. In 'taking a leap of faith', both IPs and surrogates participate in practices that are beneficial to all parties involved. This particular informal rule mobilises the affective context of online spaces and how this environment constitutes contemporary *shadow-legalities* within the sphere of surrogacy.

Rule Three: Navigating 'Socially Acceptable' Norms Around Exclusivity

> Every day [I spoke with my IP]. It was midnight, I'm still in bed texting. [My husband] tells me, 'Oh my gosh, you've got two marriages.' I'm like, 'Yeah, I do, yep.' It was really ingrained. You go looking for someone that you can see this long-term relationship with, and it becomes like a second marriage'. (Nora)

Although some of my participants refer to the 'matching process' and beyond as having several possible 'ways' to happen, there are specific ideas about 'socially acceptable norms' and appropriate behaviour that are assumed. Two related norms are the expectation of transparency around

10 Surrogacy and the Informal Rulebook for Making Kin…

and exclusivity to current or previous matches. When members hide any information from their potential matches, then it can make for awkward situations. For example, if people do not heed the unspoken rule of not dating anyone else at the same time that they are getting to know their potential match. To focus on one relationship demonstrates that people are capable of commitment and taking it seriously. For Lola this meant:

> Now we're exclusive kind of thing (laughs). I'm not looking for other people, for now, and I think if you were to try and say that straight up, say you're in a pool of people I'm looking at [it would be awkward].

Lola's views on exclusivity also extended somewhat to feelings akin to cheating when she offered to be a surrogate for her second couple. She did not want her first IPs to take offence, even though their journey together had ended a few years prior. Similarly, Celeste, an IP who had two failed attempts with her final two embryos with her gestational surrogate before seeking a traditional surrogate, wanted to talk with the surrogate who carried their first baby, and the second one with whom she could not conceive, before she wrote about it on the forum. Not all IPs and surrogates will feel this way, but it was a common expectation that no one would surprise or shock anyone they have previously embarked on a surrogacy journey with by publicly announcing a new relationship. Here we see how the language of dating is also tied up with the language of cheating and is not a consideration only for post-relationship contexts.

The idea of exclusivity is underpinned by the hope that this potential match could be 'the one/s'. This does not mean that it is a practice followed 100% of the time by everyone, but 'serial daters' were generally considered inappropriate by many of my participants. Heather also mentioned the bad manners of IPs trying to match with a surrogate whilst having personal fertility treatment. '[When IPs are] still trying other fertility avenues—[it] means they are not committed to only surrogacy/donor'. This reflects the expectation that IPs will enter the forum with both the intention to commit to surrogacy and follow the established rules of engagement and etiquette. I found from talking to several surrogates that when they felt they could trust their IPs, they really committed to wanting to help them. For example, Heather began her journey with her first IPs as a gestational surrogate but offered to be a traditional one when this route was unsuccessful. Madison also prioritised the relationship and commitment to her IPs over the type of surrogate she was willing to be:

> I was happy to do whatever method was needed but it was more the matchup that was important. I think it's more important that you get on with your IPs and that it is a proper match. It is quite an intense relationship when you're trying to conceive. It was more like looking for a couple that I got on with rather than what a couple needed… [and] finding someone that I could trust that they would look after me [during the pregnancy].

Exclusivity is more than worrying about hurting previous matches but acknowledging the surrogacy children that are born. Heather highlighted the seriousness of relationships, claiming that in comparison to dating for romance, in surrogacy 'the stakes feel much higher for me. A child is involved and I have to have supreme trust that it works out for that child—whereas dating, it's myself and my feelings and if there is a child from a relationship it is in my care—with surrogacy the child is not'. For those who create family via traditional surrogacy, being transparent with previous and potential third-party reproductive assistance means alerting everyone to the biogenetic kinship links that are being created across families. Even if the adults involved do not see them as important, the children might. Nora, a traditional surrogate for four families, told me about the relationship breakdown with one of her sets of IPs. A key issue was that the IPs did not want to acknowledge Nora's children as siblings, denying her requests for some kind of recognition that they agreed to before and during the pregnancy.

Rule Four: Mitigating Risk Through (Informal) Legal Contracts

> The emotional connection surrogates feel or wish to feel with their couple is linked to the fact that surrogacy involves a giving of oneself that in the modern Western cultural context is appropriate only in loving personal relationships. (Berend, 2012, p. 926)

A lot of these rules (although, not all) that engender forms of *shadow-legalities* around surrogacy draw from dating because they involve familiar protocols and rituals. Where rule four diverges from these is through the creation of an informal legal contract, referred to as 'letter of intent', before any assisted fertility begins, to mitigate risk. In this section I describe the way this non-legally binding document is used as a practical method to cultivate trust and establish expectations and boundaries. Regardless of whether a journey is a first, or one of many, there is anxiety inherent in each match with new people, and a letter of intent can help ease this.

On the forum, during the 'getting to know one another' period or soon after deciding to embark on a journey together, people are encouraged to write a letter of intent, that allows all parties to cement everyone's expectations. Often based on a template that members can access from the forum website, it includes guidelines to the complex questions around how long they will try to become pregnant before giving up, the topic of congenital defects and abortion, and who will have the guardianship of the baby if the IPs die. Oftentimes, this can be done in the presence of a lawyer, and even though it cannot be enforced, it appears to serve as a source of security for everyone. In gestational surrogacy journeys, parties have a mandatory obligation to discuss their intentions with fertility counsellors, both separately and then together. However, the physical letter of intent is written by IPs and surrogates in both gestational and traditional surrogacy relationships. For Heather, it was an opportunity to have frank conversations with one another: 'I guess it's [also] to check they are on the same wavelength on big issues and that it is a way of memorialising that conversation'.

As a childless traditional surrogate, it was important to Joy that the letter be drawn up before any home inseminations took place because she wanted it to also function as a plan of sorts:

> Joy: But you know, Pete, Sally and I, before we did those first inseminations, we talked a lot, and we nailed down the letter of intent, and we got ourselves all on the same page as to what we would and wouldn't do, how long we'd try for, what we were prepared… You know?
> Hannah: And what were some of those things that you decided?

> Joy: We decided that we would try for six months, so once a month for six months, and at the end of that six months either party could change their mind. We also talked about termination, because it was a big thing.

Similarly, Beth, a gestational surrogate, found it helpful as a map of sorts and reference point to refer to if needed.

> [We] sat down and wrote what we called a 'letter of intent'. So, we actually had it in writing the things we agreed on [even though] we didn't need any of it in the end' (Beth).

Conversely, Lola thought creating a letter of intent was pointless. 'I always thought that as it's not legally binding, what's the point? We talked about everything. You've just got to trust the other party 100% either way'. Instead, Lola had frank conversations when her first IPs initially visited her and her family. Even though the letter of intent is not legally binding, it holds value for those who choose to write one. For Joy and Beth, it was a visual reminder of the collective decisions they made with their IPs. It is a process that has the possibility to establish greater trust, as well as an opportunity to walk away before any fertility treatment or insemination procedures begin. It can also create a sense of control over the uncertain aspects of assisted reproduction, so as IPs and surrogates journeyed through the process, they could refer to their plan.

Berend and Guerzoni (2019), Berk (2015), and Jacobson (2016) all argue that both surrogates and IPs appreciate the legally binding (in this context) contract as a form of protection and a positive way of documenting mutual expectations. Although these scholars discuss commercial contexts, it was beneficial to all, and by mimicking a legal document, New Zealand surrogates and IPs experience similar feelings of comfort. In addition, the letter of intent in New Zealand and legal contracts in the US are discussed and refined on surrogacy forums in response to ongoing experiences (Berend & Guerzoni, 2019). In my research, not everyone felt the need for the letter of intent and found it was enough to verbally share expectations for the future. Lola's logic to not write such a letter reinforces the idea that regardless of how you establish trust, it remains fundamental to every step of the journey thereafter.

Despite not being a legally non-binding document, signing the letter of intent itself in the presence of a lawyer highlights the formality by which all parties approach it. The letter is a useful example of how a shadow of a legal contract has similar value as a legally enforceable one. As much as it gives people an opportunity to ask difficult questions and establish boundaries with what they are (not) willing to do, it can help them to identify what may be beneficial to ask or write down. Certain topics are difficult to broach, such as when to stop trying or who gets to make the decision around abortion. For example, while most surrogates believe the ultimate decision lies with the IPs, some may not agree to abort a foetus. One traditional surrogate told me that she was not religious, but she would never have an abortion and believed that the IPs signed up for a baby, regardless of whether they test for a genetic condition. Ultimately, in New Zealand, it is up to a pregnant woman to make that decision, surrogate or not, demonstrating the importance of IPs and surrogates aligning in values during the matching process.

Understanding Surrogacy Shadow-Legalities: Informal Rules and Kin-Making

These various informal rules guide and shape the lives and behaviours of members within the surrogacy community—sometimes mimicking informal online dating rules and at other times approximate formal legal documentation mimicking marriage contracts. As I have proposed in this chapter, these informal rules operate as *shadow-legalities*, as different processes members can rely on to help them make sense of a practice that they have no reference for historically. These laws are not extensive of every informal rule but nevertheless reveal nuances of the ways (un)desirable behaviour is regulated by more senior members within the community. These *shadow-legalities* also refer to reproduction that is practised in the shadows of conventional heteronormative conception and thus require creativity to successfully navigate. They also exist in conjunction with, or because of, ambiguous laws around surrogacy. With no guidebook to follow, senior members in the surrogacy community have established explicit and implicit codes of conduct within which relationships are carefully crafted and cared for and are sometimes only apparent when things go wrong.

The four shadow rules or laws I have identified in this chapter are encouraged and reinforced where relevant by senior members. Rule number one, that everyone is expected to follow the forum protocol to remain members of the MBS forum, access support, and find a potential match, is underpinned by the assumption that *if* people stay within these specific parameters set out by moderators, they too may be successful. This functions as a visible pathway for new members and people have a clear idea of what is (un)acceptable behaviour. Those who do not engage with others' stories or build rapport are much less likely to find a match because potential matches do not see what their personality is like or if they want to get to know them more. The more specific rules around what constitutes inappropriate behaviour are considered more grievous, and moderators can enforce these by giving a warning to the member or removing them. The way all my participants referred to and reiterated the forum rules, even if they struggled with them like Christina, demonstrates how they become internalised. Conversely, Christina's experience shows that having instructive rules can be anxiety-provoking and may not always helpful. Not being able to put in the time required to be interactive and open does not necessarily mean people are not worthy of finding a match.

Rule two, taking a leap of faith and trusting fate, is full of romantic imagery and hope about building a relationship with others. This rule is more complex than rule one because it is based on intuition that may or may not lead to a positive experience. I identify it as an informal rule because it is as equally embedded in my conversations with surrogates and IPs alongside more overt rules. It almost subverts the notion of a rule because it encourages members to ultimately heed their own intuition and heart when making decisions. It must 'feel' right. We can see this in Madison's case where she had pragmatic criteria but offered to be a surrogate to her IPs on the first date because it was 'love at first sight'. Equally, Tracey chose Louise and Mike because she felt it was right even though her initial criteria was to gestate a baby for a younger couple. Rather than dismiss pragmatism altogether, this rule reminds members that aligning values works well in conjunction with emotional attraction.

Rule three is more about the socially acceptable norms people are encouraged to follow. Exclusivity is an important norm and relates to dating one potential match at a time rather than serial dating several people at once. If IPs are not transparent and are undertaking other fertility treatment at the same time, it is deemed disrespectful especially because surrogates are in high demand and want to help people who have exhausted all other avenues. Potentially, because kin ties are made via genetic or social connections during the surrogacy arrangement process, some of my participants emphasised the importance of talking with their previous matches before they embarked on another journey with a new person/s. This was seen as a mark of respect. Vulnerability can manifest in different forms, and these norms seem to be in place to avoid hurting others as much as possible. As mentioned in regard to rule one, people are more likely to turn to the community for emotional support because they do not feel like their friends or family would understand their personal predicament, given how different surrogacy is to their experiences.

Rule four, the creation of a letter of intent is an informal contract by those who utilise it. All members can access a template that is available on the forum website and can ask others what they prioritised. It may not always be followed, but it is encouraged because it forces people to talk about difficult topics, ask questions, help establish boundaries and have a clear plan in place if difficulties arise. Signing it in front of a lawyer, even though it is not legally enforceable, still works to help formalise everyone's intentions. Ultimately, it is used to identify and mitigate problems before they (may or may not) happen.

Conclusion

This chapter focuses on traditional and gestational surrogacy arrangements in New Zealand. I have introduced the concept of *shadow-legalities* to frame discussion about the informal laws and processes that shape the way people approach finding a match and form their relationships with one another in order to make kin in non-traditional ways. Following Riggs and Due (2013), people who practise surrogacy are reproductively vulnerable because they are not creating kin via heteronormative reproduction. They rely on the informal rules they co-constitute because there is no guidebook or reference beyond what others have experienced and learnt along the way, positive and negative. When new members see clear pathways to success, they may feel more in control. Unfortunately, as shown in Christina's narrative, the pressure to conform and engage intensely with the forum was too stressful. Within Euro-American kinship ideology, narratives of love, dating, commitment, compatibility, and choice are often seen as essential prerequisites to legally recognised unions, whether marriage or de-facto partnerships.

Because assisted reproduction is traditionally conceptualised as a heteronormative space where affection and love is situated, it is not surprising that members of the surrogacy community in New Zealand rely on familiar terminology of romance and dating to contextualise their experiences during both the period of searching for a surrogate or IP and the ensuing relationship that develops. Whether finding a match, mid-journey or after, members find it helpful to draw on their prior dating experience or knowledge to guide their behaviour in an unfamiliar situation. They transpose and reshape their knowledge of other informally constructed spaces—like online dating—into the surrogacy community and it ultimately helps them to feel in control. Although Heather emphasised that people should make up their own minds about who to choose as a surrogate or IP, many members do make decisions or act based on how potential or actual matches pan out on the forum and their behaviour offline. In this sense, the informal rules provide guidelines that facilitate kin-making and reproduction in non-traditional ways. The *shadow-legalities*, such as the letter of intent, serve as quasi-formal rules that dismantle heteronormative privileges and claims to reproduction and kin-making. Ultimately, as the experiences of my research participants show, it is the rules that make the experience of assisted reproduction via traditional surrogacy and gestational surrogacy possible.

Notes

1. Intended parents are known in the surrogacy community as simply 'IPs'; a descriptor for everyone who has sought, or currently seeks, a surrogate.
2. Romance in the traditional sense or sexual relations does not take place.

References

Berend, Z. (2012). The romance of surrogacy. *Sociological Forum, 27*(4), 913–936.
Berend, Z. (2016). *The Online world of surrogacy*. Berghahn Books.
Berend, Z., & Guerzoni, C. S. (2019). Reshaping relatedness? The case of US surrogacy. *Antropologia, 6*(2), 83–100.
Berk, H. L. (2015). The legalization of emotion: Managing risk by managing feelings in contracts for surrogate labor. *Law & Society Review, 49*(1), 143–177.
Ferguson, J. (2006). *Global shadows: Africa in the neoliberal world order*. Duke University Press.
Galemba, R. B. (2008). Informal and illicit entrepreneurs: Fighting for a place in the neoliberal economic order. *Anthropology of Work Review, 29*(2), 19–25.
Gibbs, J. L., Ellison, N. B., & Lai, C.-H. (2011). First comes love, then comes Google: An investigation of uncertainty reduction strategies and self-disclosure in online dating. *Communication Research, 38*(1), 70–100.
Goodfellow, A. (2015). *Gay fathers, their children, and the making of kinship*. Fordham University Press.
Helena, R. (1994). *Surrogate Motherhood. Conception in the Heart*. Boulder, West View Press.
Jacobson, H. (2016). *Labor of love: Gestational surrogacy and the work of making babies*. Rutgers University Press.
May, A., & Tenzek, K. (2016). "A Gift We Are Unable to Create Ourselves": Uncertainty reduction in online classified ads posted by gay men pursuing surrogacy. *Journal of GLBT Family Studies, 12*(5), 430–450.
Nordstrom, C. (2000). Shadows and sovereigns. *Theory, Culture, and Society, 17*(4), 35–54.
Riggs, D. W., & Due, C. (2013). Representations of reproductive citizenship and vulnerability in media reports of offshore surrogacy. *Citizenship Studies, 17*(8), 956–969.
Sampson, S. (2003). "Trouble Spots": Projects, bandits, and state fragmentation. In J. Friedman (Ed.), *Globalization, the state, and violence* (pp. 309–342). AltaMira Press.
Shaw, R. (2008). Rethinking reproductive gifts as body projects. *Sociology, 42*(1), 11–28.
Shaw, R. M. (2020). Should surrogate pregnancy arrangements be enforceable in Aotearoa New Zealand? *Policy Quarterly, 16*(1), 18–25.
Teman, E. (2010). *Birthing a mother: The surrogate body and the pregnant self*. University of California Press.

11

Constructing Gay Fatherhood in Known Donor-Lesbian Reproduction: 'We get to live that life, we get to be parents'

Nicola Surtees

Introduction

Assisted reproductive technologies contribute to increasingly complex familial futures in Aotearoa New Zealand (hereafter, Aotearoa). When known donors provide sperm to prospective parents the status of their social relationships to the children they help to conceive may be uncertain, because they have no obvious place within kinship systems. While these donors will be familiar to prospective parents as relatives, friends, acquaintances, or people met on social media platforms, figuring out where they fit in the lives of resulting families occurs without an established script for managing this process.

Use of known donors in assisted reproduction in Aotearoa is encouraged by Fertility Associates, the largest fertility service provider, on the

N. Surtees (✉)
University of Canterbury, Christchurch, New Zealand
e-mail: nicola.surtees@canterbury.ac.nz

© The Author(s), under exclusive license to Springer Nature Singapore Pte Ltd. 2022
R. M. Shaw (ed.), *Reproductive Citizenship*, Health, Technology and Society,
https://doi.org/10.1007/978-981-16-9451-6_11

basis that they enable timely initiation of clinic mediated conception procedures (Fertility Associates, n.d.). An estimated 60% of women accessing this provider choose to recruit a known donor to work with them through the clinic, rather than wait for an identity-release donor to become available (Chisholm, 2016). Currently, the provider advises the wait for an identity-release donor is approximately 18–24 months. Identity-release donors, the only legislated alternative to a known donor using standard clinic mediated conception strategies, are donors whose identity is unknown to parents at the time of donation but who can become known to them and their children in the future under the provisions of the Human Assisted Reproductive Technology Act 2004 (HART Act).

The emphasis on known donors as one solution to fertility clinic waiting lists occurs against the background of societal pressures for fathers to be known biogenetically and socially. Gay men who become known donors for lesbian women because they want to be recognised as fathers can press the advantage afforded by the dominant discourse that all children have the right to a father and/or information about their paternal origins when negotiating reproductive relationships. Who these men think they are in relation to expected or actual children has, however, received scant research attention to date (but see, Côté & Lavoie, 2019; Dempsey, 2012a, 2012b; Riggs, 2008a, 2008b). What we do know is that 'the relation between donors and offspring is highly contested' (Griffin, 2020, p. 141).

This chapter draws from a qualitative interview study in Aotearoa that explored the negotiation of relatedness in known donor-lesbian reproduction (Surtees, 2017). Sixty men and women, across 21 different known donor-lesbian familial configurations at various stages of forming families through known donor insemination participated in the study. The chapter focuses on some of the ways in which gay couples exercise agency and choice, through their shared experiences of being, or planning to be, known donors for lesbian couples and what they do to construct themselves as joint fathers *and* parents. Understanding distinctions between what it means for a gay known donor and his partner to be fathers and parents in a context where lesbians typically position known donors as fathers—but not parents—or as other flexibly defined male figures such as uncles or friends (see, e.g., Hayman et al., 2014; Kelly, 2011; Nordqvist, 2012; Surtees, 2011) contributes to knowledge about possibilities for fathering/parenting identities and practices. The chapter argues that the narrative construction of fathering/parenting identities, and the fathering/parenting practices imagined or sustained, reconfigures notions of what it means to be a father/parent while simultaneously reinforcing traditional meanings.

Changing Kinship Narratives: A Sociology of Personal Life

Eliciting the participants' stories about known donor-lesbian reproduction for the study involved little effort during interviews. Spontaneous prefacing comments such as 'there's a story coming up' were frequent. The field of kinship studies provided a framework to make theoretical sense of these stories. In conventional kinship thinking, kinship is viewed as an unchangeable fact of nature (Carsten, 2004), whereas 'new' kinship thinking has been attributed with reformulating kinship in the direction of relatedness (Mason, 2011). While assumptions pivotal to old kinship thinking emerged in the participants' stories, including the assumption that biogenetic connections produce 'fixed' relationships (Nordqvist, 2019a), their stories remain profoundly relational narratives that include relational content and descriptions of relational practices. This section of the chapter therefore focuses on a sociology of personal life, as one example of the move towards relationality evident in contemporary kinship discourse.

Closely associated with the work of Carol Smart (2007, 2011), personal life's core emphasis on relationality highlights the ways in which people's everyday lives are relational in character. Relationality, Smart (2011) states, 'conjures up the image of people existing within intentional, thoughtful networks which they actively sustain, maintain or allow to atrophy' (p. 17). In stressing relationality, personal life broadens established kinship boundaries, allowing new analytical approaches for understanding the importance of connectedness with others to gain in significance (Smart, 2007). As Nordqvist (2019b) maintains, because questions about connectedness feature in forms of assisted reproduction that utilise gamete donation, this perspective can capture key facets of the donating experience that are currently under-researched. In this study, personal life proved invaluable to this end. In particular, personal life captured the ways in which the participants—as parties to the donating experience—are embedded within and constituted by a field of persistent or 'sticky' (Smart, 2007) adult-adult and adult-child relationships. In refusing to take family as the only reference point for relationships (Smart, 2007), personal life opened conceptual spaces suited to exploring the participants' divergent relational narratives; both family relationships *and* different forms of relatedness, including kin-like relationships and non-kin relationships needed to be accounted for, without presuming what forms particular familial configurations took or how they were distributed across households. Personal life's ability to make room for wide-ranging forms of relatedness and the ways these shift and move over places and spaces with no one form, place, or space privileged over another was therefore instrumental to analysis (Smart, 2007).

Moreover, in reaching beyond established kinship boundaries, personal life facilitates a focus on kinship as a process, something people actively negotiate and do in everyday life that is specific to them and their relationships (Mason, 2011; Nordqvist, 2019a). Mason (2011) elaborates: 'many people get involved in the process of *working out* who they are related to, rather than being able always to take this for granted as a given' (p. 65, italics in original). Working out new ways of relating was a continuing exercise for the study participants, including the men introduced in this chapter.

Issues of individual agency and choice can also be considered from a personal life lens. As Smart (2007) observes, 'to live a personal life is to have agency and to make choices, but the personhood implicit in the concept requires the presence of others to respond to and to contextualize those actions and choices' (p. 28). Personhood, or the development of self and identity then, is determined relationally, and emerges through interactional processes as relationships are formed, sustained, and dissolved (Finch & Mason, 2000; Smart, 2007, 2011). Such conceptualisations of the self and identity are frequently obscured in Western thought through a stress on the bounded and boundaried individual. This is an individual who might seek out relationships but who could equally well live independently of others, someone who exercises free agency and is solely responsible for his or her own choices (Smart, 2007). Mason (2004) contends that the purchase of individualisation theses, persuasive in sociological explanations of social change in the global North, is at odds with such conceptualisations. They can be countered by empirical analyses that foreground the role of social connectivity in self and identity construction across diverse contexts and scenarios. Rather than simply dismiss the relevance of individualistic discourses and practices to self and identity construction, she advocates attention to whether and how such discourses and practices are intertwined with relational discourses and practices. While traces of individualistic tendencies are evident in some of the participants' stories, they are, as noted earlier, profoundly relational narratives about relational connections. Illuminating the ways individual agency is situated within sets of relations, the narratives foreground the existence of others who must be taken into account and responded to. Similarly, they foreground the ways in which choices are made with regard for the needs and feelings of these others.

In sum, stories, as one example of interactional processes, are produced within relationships. As illustrated in this chapter, participants used their stories to narratively construct particular selves and identities. Paying attention to their narratives of embeddedness—the ways in which they are located within intentional, thoughtful networks and the 'stickiness' of their bonds (Smart, 2007)—highlighted what they thought and felt about the kinds of relationships they had established or imagined establishing with one another and children. Brockmeier and Carbaugh's (2001) observation is apt: 'The stories we tell ourselves about ourselves and others organize our sense of who we are, who others are, and how we are to be related' (p. 10).

Study Methods

As previously indicated, the study this chapter draws from investigated how the parties to known donor-lesbian reproduction negotiate relatedness. The 60 adults who participated in the study included gay and heterosexual known donors, male and female partners of donors, and lesbian couples across a range of different familial configurations. These adults had all either previously become parents through known donor insemination, were actively pursuing conception, or planning future parenthood using this method.

Initial recruitment of participants focused on gay men and lesbians who met these criteria. Promotion of the study was therefore directed towards gay and lesbian-targeted national and regional organisations and social groups. As recruitment proceeded, the criteria broadened to include familial configurations inclusive of gay men, lesbians, *and* heterosexual men and their partners. It had become apparent that it was not the participants' sexuality that mattered per se. What mattered was their potential to provide insight into a wide range of social identity possibilities and roles for gay or heterosexual men as known donors for lesbians, and those of their partners, in relation to children, given their uncertain location within the kinship structures put around the children.

Participant stories about known donor insemination were collected through narrative interviews; in opening up topics and accommodating long accounts, this method provides the conditions necessary for storytelling (Bold, 2012). Because relatedness must be negotiated together, the adults were invited to share their stories about this process in their familial configuration groupings where practically possible.

Twenty-six semi-structured narrative interviews were subsequently conducted: 10 group interviews with from 3 to 5 members of particular familial configurations, 11 couple interviews and 5 individual interviews. The interviews were digitally recorded and transcribed before being returned to the adults to check for accuracy. The ethnographically rich interview data captured the divergent ways they storied their experiences of negotiating relationships.

There are numerous possibilities for narrative analysis. A storied approach focused on a thematic analysis of what participants had to say about their experiences rather than an analysis of how they went about the telling of these experiences was adopted (Riessman, 2008; Sparkes, 2005). In other words, the content of what was said was attended to, not the form with which it was said, or the actual structures of speech or social processes that were used to say it. The study was granted ethics approval from the University of Canterbury New Zealand Human Ethics Committee (HEC 2009/158). Ethical requirements were adhered to, such as the protection of participant identities. Real names were not used in the study or resulting publications, including in this chapter.

Identity Narratives: The Construction of Fathering and Parenting Selves

Three gay couples are introduced in this section of the chapter. The men use their stories about known donor-lesbian reproduction to narratively construct themselves as fathers and parents. In two cases, the men's stories include reflections from the lesbian couples they collaborated with.

As part of the relatively new trend in planned gay fatherhood/parenthood (Carneiro et al., 2017), the first two couples engaged in determined efforts to become fathers/parents through initiating the recruitment of lesbian couples prepared to participate in co-parenting arrangements with them. These kinds of arrangements between gay men and lesbians, characterised by the intention of each of the adults to take up a parental role of some description (Surtees & Bremner, 2020), provide a relatively accessible biogenetic pathway to gay fatherhood/parenthood in comparison to traditional or gestational surrogacy (for recent studies of such arrangements see, Côté & Lavoie, 2019; Herbrand, 2017). They may also be easier to achieve than social pathways such as fostering and adoption, particularly given these pathways can be cost prohibitive and/or may not be legally available in some jurisdictions (Park et al., 2015).

While active in the matter of becoming fathers/parents, these two couples articulate a subordinate status in family-making processes relative to the women they either imagine collaborating with or actually collaborate with, consistent with studies that note gay fathers/parents are usually viewed as secondary parents (Andreasson & Johansson, 2017). This occurred as they projected ahead to possible co-parenting arrangements or in early negotiations of them.

Rather than engage in the kinds of deliberate attempts to become fathers/parents that were a hallmark of the previous two couples' journeys towards fatherhood/parenthood, the last gay couple were drafted in as a necessary third party to a lesbian couple's conception plans. The men's articulation of a subordinate status in relation to them, both in initial negotiations and on a day-by-day basis following the birth of a child, was therefore unsurprising.

Kole and Fraser

Kole and Fraser, who both grew up in Eastern Europe, met in their early twenties before relocating to Aotearoa. Aspiring fathers and parents, they undertook an online search for a lesbian couple interested in a co-parenting arrangement. As Fraser said, co-parenting was their 'ideal scenario.' With each man intending to be positioned as a father/parent, their family would incorporate biogenetic and social fathers/parents, biogenetic and social mothers/parents and one or more children, whose biogenetic father/mother would not be connected by a sexual relationship. It would also incorporate two legal parent-child relationships.[1]

Kole and Fraser's preferred scenario was influenced by old-fashioned ideas about the necessity of opposite sex parents. Fraser explained, 'We just like this option because it would give our kids mothers and fathers, so they don't grow up feeling that they've already lost something.' Kole added, 'The missing parent is a huge issue… it's like a grief. It's extra baggage for the kids.' The men use their narrative to construct themselves as personally responsible fathers/parents to be, eager to shield their future children from the problems that could arise if they failed to provide them with continuing contact with both male and female parents.

Assumptions about the significance of biogenetic relatedness also influenced the men:

Fraser: It would be great if we were to have like a genetic link.
Kole: We would like to be biological fathers. Well, at least first. I think it doesn't matter after you bond with your kid. It can be not related to you, and you don't care. But, at the moment, I feel I would like to be biologically related…
Fraser: The biological link is not as important to me, but obviously I wouldn't mind if there was some genetic relation…
Kole: It just feels different at the moment. But you know, probably it doesn't make any difference.
Fraser: Yeah—if I think about it, it doesn't matter. If you raised that child, it's yours. You can have good genes or bad genes, and anything can turn out at the end even if it's your genes or not.
Kole: It's not because of the genes. I just think it could give some kind of extra connection. I don't know. I have no idea!

In this extract, both men make contradictory statements about the significance of biogenetic relatedness, something observed in other studies of gay fatherhood (see, e.g., Nebeling Petersen, 2018). Ultimately, they accept that biogenetic fatherhood/parenthood is not as important as a social relationship to a child they might parent. They argue that biology *doesn't matter* because, *if you raised that child it's yours*. Only one of the men will donate sperm. Given the other has no option but to take up social fatherhood/parenthood, the idea that there are alternative relational bases for fathering/parenting is useful for them, because it validates social possibilities for fatherhood/parenthood.

Kole and Fraser explained their vision for co-parenting arrangements with a lesbian couple:

Fraser: Let's start with the best-case. It is really like a 50–50 [equal time split across the couples' homes]. Or, we were thinking another arrangement could include—
Kole: Two kids.
Fraser: Two kids: one for each of us. One at our place [fulltime], and one at their place [fulltime].
Kole: Because we don't want to be just weekend dads.
Fraser: No.

Kole: That's not really enough. I think we cannot really be a part of the kid's life if we only see him on the weekends—that's like uncles. It's different... The kid's everyday life and important decisions, we would not always be a part of because we'd just be there at the weekend. No—and I really would like to do the nasty parts: changing nappies, burping the baby.

While the men exercise agency in their pursuit of fatherhood/parenthood, hoping for a co-parenting arrangement with a lesbian couple based on an equal, cross home time split, the idealism imbuing this option is tempered by anticipated constraints on what might actually be possible. In recognising they may have to accept the levels of involvement the lesbian couple define they articulate a subordinate status. As Fraser said, 'It all depends on the other couple, what we can really arrange.' For this reason, they had decided they were willing to accept becoming weekend dads/uncles for a *first kid* as a *test try*:

Kole: Well, we thought about this and it would be okay.
Fraser: Yeah, we would go for it.
Kole: For our first kid and then we'll see—because, why not? Why not? Why not make a lesbian couple happy by giving them a child?
Fraser: And we can be a part—
Kole: For a small amount. It'll be like a test try...
Fraser: It's just that we really want to have that full experience of parenting. And the other thing is that, if you really work hard with the couple then later, the situation can maybe change—you never know.
Kole: Maybe they can let you more in to that.

In these extracts, Kole and Fraser use their narrative to construct themselves as new fathers—emotionally responsive, competent, and equal caretakers (Stevens, 2015). Simultaneously, they draw on discourses of participatory fatherhood and involved father divorce discourse as resources to make sense of possibilities for their co-parenting involvement. Becoming *weekend dads*—likened to being uncles in the first extract and to a *test try* in the second extract—is not consistent with their aspirations for parenthood—this relationship to the planned child is *not really enough*. They are aware that the relational statuses of weekend dads/uncles may not afford opportunities to *really be a part of the kid's life*, to help make *important decisions*, or to participate in the full range of caring practices, including *the nasty parts*—in other words, to engage in *that full experience of parenting*. Arguably, nappy changing—one of *the nasty parts*—is a symbol of involvement in the nitty gritty practical tasks of parenting a small child. Such practical tasks, typically synonymous with activities of mothering, are ones that men have been identified as avoiding. By seeking to embrace these kinds of tasks, Kole and Fraser construct themselves as different sorts of fathers/parents—potential mothering male parents—while distancing themselves from dominant hegemonic masculinities that frame fathering/parenting in conventional ways. In their view, fathering/parenting relationships to children are understood as flexible, negotiable and centred on practices of involvement. Fatherhood/parenthood is conferred through extensive involvement, not biogenetic relatedness, or some form of secondary role, which would only serve to reinforce their subordinate status.

Just as divorcing or separating heterosexual men must start planning for fathering/parenting involvement in response to the dispersal of parenting across new households, Kole and Fraser must actively plan for their fathering/parenting involvement using what they know is sometimes the outcome for heterosexual men in the divorce or separation context—models of the weekend dad/uncle—to shape what they consider acceptable. Unlike divorcing/separating men, whose parenting relationships are sustained when an intimate relationship has broken down, resolving issues about how they will care for their child and where he or she will live will not be complicated in the same way (Donovan, 2000; Dunne, 2000). A willingness to *work hard* at couple-couple relationships is nevertheless considered important.

Kole and Fraser's narrative articulates their aspiration for emotionally engaged fathering/parenting. They anticipate inventing new kinds of family lives together with a lesbian couple, while simultaneously relying on established heteronormative conventions as resources in their stories, including old ideas about the significance of opposite sex parents for children and those generated by divorce and separation politics and practices. While their idealism was yet to be practically tested, this was not wholly the case for the couple at the centre of the next narrative.

Wilson and Johan

In his late twenties Wilson partnered with Johan, a similar aged peer.[2] The couple married in Wilson's homeland of Canada before settling in Aotearoa. Once settled, they decided to explore becoming fathers/parents. Like Kole and Fraser, their preferred pathway to fatherhood/parenthood was a co-parenting arrangement with a lesbian couple. Vivian and Moira, who they met online, were keen to collaborate with them.

Wilson expected to provide the sperm for Vivian to conceive. When he subsequently learned of his suboptimal fertility, he concluded Johan should donate in his place. From Vivian and Moira's perspective however, further negotiation was warranted about which of the men should act as the donor and become a biogenetic father/parent, and which a social father/parent, as the other's partner:

Moira: The first interesting test… came when we found out Wilson's swimmers weren't really active.
Wilson: Yeah… And now even Johan's—they're [the fertility service staff] saying they want to improve even Johan's fertility because it's threshold… What are these guys? Olympic swimmers or what?
Moira: Our approach is quite interesting. We kept wanting to ensure that this switch, of who is going to be the biological dad, that the boys were comfortable in their relationship with it. So, we kept going subtly back: 'You know, there's still a chance that Wilson can be the dad here. It just means we're going to have to work at it…' We wanted to make sure that there wasn't any weird dynamics.
Nicola: So, it wasn't just about which sperm swam fastest?

Moira: We didn't want an alpha male defined by science—right: 'You're going to be the dad cause you're the fitter. The better-bred winner.' We wanted in their relationship, for them to be very sure that there wasn't going to be any dynamics about who is going to be the biological father in their relationship that might then affect our relationship that then gets displaced on to the child…

Wilson: To be honest, I just thought this is the most logical process… I know it is very scientist of me but… this makes the most sense.

Wilson deflects the focus on his fertility Moira opens the exchange with by drawing attention to Johan's fertility. Possibly, he feels vulnerable. As Griffin (2020) states, 'virility and sexual prowess are laid open to inspection in the process of the intimate labour of sperm donation, rendering the donor vulnerable on multiple fronts' (p. 139). At the same time as using his story to construct himself as reasonable, Wilson intimates any expectation that sperm should be *Olympic swimmers* is unreasonable and by implication, constructs the fertility service staff as unreasonable. Because he projects himself as someone who rejects constructions of masculinity that are bound up with hyper fertility, the women are eventually able to accept Johan as an alternative prospect for biogenetic fatherhood/parenthood.

Wilson also deflects a focus on his fertility by deploying a scientific discourse as a narrative resource. His *very scientist* logical approach to conception communicates a science-orientated, pragmatic self—someone concerned with the facts of his reduced sperm motility and how to address this, rather than someone with feelings about the status of his sperm. He implies the men's sperm are just a means to an end and as such it does not really matter which of them provides it and subsequently claims biogenetic fatherhood/parenthood. His refusal to accept that a biogenetic contribution to conception is the only means by which fatherhood/parenthood can be conferred is useful for him because it will support his social fathering/parenting, the only option available for him at this point.

The couples agreed Vivian and Moira would take main responsibility for their future child's care in their home as his or her legal parents and guardians. Beyond this, little practical detail about the men's involvement had been decided. As Wilson said:

> It is hard to predict how much we're going to be involved… We have to be careful not to overly prescribe the situation before it happens. I keep coming back to that idea that there are some things that need to be rigorously negotiated and there are other things that need to be left to chance.

Despite some uncertainties, the men anticipated that their fatherhood/parenthood could be conferred without the kinds of extensive involvement that Kole and Fraser considered necessary to fully experience parenting. While they were open to different possibilities, they were not seeking a co-parenting arrangement based on an equal, cross home time split. Rather, as Wilson explained, they would do the moving:

> It's about us moving around more than the child… I think they should be exposed to our home, so they know it's their home, but I don't believe in 'pass the baby'… Maybe when the child is a bit more able to—I think it can work. But I think as a newborn it would be terrible to kind of move it around… I get really disappointed when I see heterosexual couples who divorce—they put all the pain onto the child. 'You will move to these houses, these times, with these people.' I think that's terrible because it is your relationship that has fallen apart.

Divorce discourse informs Wilson's opinion about possibilities for parenting across residences. While Kole and Fraser utilised divorce discourse as a resource to reinforce their parenting status and participation in care practices, Wilson utilises this in ways that potentially diminish he and Johan's parenting status and possibilities for participation in these practices, as subordinate actors to the women. In the post-divorce model he draws from, children are expected to move between their parent's residences—while he wants his child to come to know their house as a second home, because he or she will not be passed backwards and forwards his/her experience of this home will be limited.

As non-residential parents, Wilson's emergent, organic approach to practical details also arguably diminishes his and Johan's parenting status and possibilities for participation in care practices. While they may have aspired to the image of the new father, in practice they did not choose to explore avenues to fatherhood/parenthood that would enable them to fulfil this image in a fulltime, residential capacity. Instead, they imagined that they would leave the daily, residential work of rearing children to Vivian and Moira. A discourse of paternal choice, positioning paternal involvement as optional, is invoked. According to Mallon (2004), 'In a family where there is a mother, a man can decide how much or how little he wants to participate' (p. 138).

Wilson and Johan's narrative emphasises some of the resources they use to construct themselves as prospective fathers/parents. As the intending social father/parent, Wilson constructed himself as a particular kind of father/parent through his behaviour and actions. He understood his doing of fathering/parenting would occur in ways unrelated to biogenetic connections and that what this looked like would be revealed over time. Unlike these men, the last couple profiled in this chapter were already fathers/parents at the time of their interview.

Max and Patrick

Believing his prospect for fatherhood/parenthood was limited Max had decided that if a lesbian couple asked him to donate sperm he would. Saying, 'I'd just always thought I'd do it', donating was to be conditional on involved fatherhood/parenthood. Rather than actively recruiting lesbian couples to participate in a co-parenting arrangement like the men in the previous two narratives, Max waited to be approached by potential couples.

Max and Patrick had been together for a month when Patrick's friends, partners Nicole and Jeannie, told the men they were looking for a donor. The women—who had previously attempted to conceive through clinic-based inseminations—expected to locate parenthood exclusively in their couple relationship in keeping with clinic processes structured by laws governing assisted reproductive procedures and parenthood. In considering Max as a donor, the women used their experience of the clinic and the dominant heteronormative model of family as resources to reinforce their position as key actors in family-making processes.

When the couples met to discuss possibilities, Max voiced his desire for contact with their future child in return for his sperm. Investing sperm with meaning, he understood the biogenetic contribution to conception it represented could be used to confer his fatherhood/parenthood. Jeannie recalled their response: 'Max wanted to see him. He wanted access to him, didn't he? We were like: "Hmm, don't know if we like that! This is a surprise. We need to think about this now."' Moreover, Max wanted to formalise his fatherhood/parenthood through the inclusion of his name on the child's birth certificate. As he said:

> I suppose at the beginning, and not necessarily knowing how things were going to work out… I wanted the child to have some recognition of me. Even if I wasn't part of his life, I thought being on his birth certificate would be a kind of lasting thing for him to be able to see, if things didn't work out.

Arguably, Max had some bargaining power, even as a subordinate actor. He exerted his influence using resources at his disposal to orientate Nicole and Jeannie away from the model of family they had previously aspired to and towards a model of family inclusive of involved fatherhood/parenthood. His knowledge of the women's circumstances was one such resource; he had learned of their conception history and continuing hope for a child and could use this to reinforce his conditions. His access to the discourse that all children have the right to and need a father and/or information about their paternal origins via the mechanism of the birth certificate was another useful resource.

While Nicole and Jeannie accepted Max's terms, doing so represented a significant change to their original plans to confine parenthood to coupledom. Max had agreed the women would be the child's primary parents providing most of his or her care in their home. While this arrangement reinscribed his status as a subordinate actor, the women's willingness to revise previously held conceptions of family to accommodate him would nevertheless alter their family-making trajectory irrevocably. It also raised vexing conceptual, relational, and pragmatic questions. With Max constituted as a father and additional parent, what form would the family now take? How would Patrick figure in the family, both as Max's partner, and as a man who had also openly declared a long-term wish to be a father/parent? Who would he be to the child? How would co-parenting across residences work in practice? Without ready answers to these kinds of questions, insemination attempts began.

Nicole quickly became pregnant; Patrick joked Max thought his sperm 'was the most potent stuff on the planet.' From the moment of Elliot's birth, through until the couples' joint interview when he was three years old, the pragmatics of everyday life with a young child served to facilitate the resolution of their earlier questions. Time, experience, and involvement saw the family consolidate as a multi-parent, cross-residential model with the constitution of Max as a father and additional parent confirmed from the outset with Patrick's social fatherhood/parenthood following later. Fluid family boundaries accommodated Max and Patrick's couple relationship, Nicole and Jeannie's couple relationship, Max and Nicole's reproductive relationship, Max and Nicole's respective biogenetic fatherhood/motherhood, Patrick and Jeannie's respective social fatherhood/motherhood, and four distinct adult-child relationships.

Time, experience, and involvement also enabled the couples to reach some shared understandings about possibilities for social relationships as an alternative relational basis for fatherhood/parenthood with respect to Patrick. Patrick's place in the family and relationship and role with Elliot were initially much less certain than was the case for Max, whose biogenetic contribution to the boy's conception allowed him to readily claim a place as his father and additional parent given prevailing discourses that conflate the two. Similarly, possibilities for Max's actual relationship and role with the boy were also accessible to him, through new ideas about options for known donor involvement in the families of lesbian couples and older ideas about the form non-residential fatherhood/parenthood can take in the divorce or separation context.

Both Patrick and Max suggested the brevity of their relationship precluded consideration of Patrick's place during early reproductive negotiations between Max and the women. As Patrick said, 'I was kind of like a spare wheel at that point.' Max intended to become a father/parent in exchange for sperm regardless of whether or not his relationship with Patrick endured over the long term. He commented: 'From my perspective, I was going to do this. Whereas Patrick was—we'd only known each other two months.' Max's comment suggests he did not assume Patrick would have a part to play in the arrangements he worked to secure with the women.

When Patrick first brought Max and the women together, he did not realise that Max was intent on fatherhood/parenthood: 'At the beginning I thought that what we were doing was that Max was providing an opportunity for Nicole and Jeannie to become parents… I thought Max was a sperm donor.' In expecting social separation between Max and the child, Patrick accessed existing ideas underpinning donor insemination for heterosexual couples consistent with clinic practice. Unlike Max, Patrick was informed by the conventional assumption dominant in many gay and lesbian circles that the provision of sperm by an alternative means to vaginal sex severs a father/parent donor-child relationship (Dempsey, 2004). In his view, being a donor was not conflated with fatherhood/parenthood.

When Patrick came to realise Max was to be a father/parent, he did not immediately see possibilities for himself to become a social father/parent as Max's partner despite wanting children. Nordqvist (2020) acknowledges that 'making kin' can be more difficult for some than others. Patrick's assumption that a biogenetic contribution to conception is necessary to the conferral of fatherhood/parenthood is in direct tension with his understanding that being a donor is not conflated with fatherhood/parenthood. Despite this, his investment in Max, and Max's investment in Elliot, led him to similarly invest in the boy. Because the men lived together, and Elliot spent regular time in their home, opportunities for Patrick to help Max father/parent Elliot emerged naturally. Time, experience, and involvement therefore marked Patrick's transition from the partner of a sperm donor, to the partner of a biogenetic father/parent through sperm donation, to his construction as a social father/parent. As Patrick said: 'It wasn't until Max and I became closer, and we started to have Elliot [to stay], that it changed for me. I thought: "Well, I'm not someone now on the periphery anymore. This is us."' While unexpected, this was a transition that was valued by each of the adults and one that exemplifies revisions to conceptualisations of family and adult-child relationships and roles.

Côté and Lavoie (2019) observe that ongoing, open-minded negotiation in co-parenting arrangements—such as Max, Patrick, and the women engaged in—is key to their success. Given the divergent expectations Max and the women initially brought to bear on their negotiations, and Patrick's sense of himself as *a spare wheel* to these negotiations, conflict may have emerged as they navigated everyday living in such a family had they not adopted this approach. Instead, the adults reported a high level of satisfaction in the relationships and roles taken up with Elliot. In Patrick's words, what emerged was 'the most remarkable situation where… we get to live that life, we get to be parents.'

Concluding Discussion

The three gay couples featured in this chapter reflexively negotiate competing ideas about who can be a father or a parent. While agency and choice play an important role in these negotiations, their stories provide convincing examples of the ways these concepts are characterised by relationality and attentiveness to others from within the context of particular constraints.

Kole and Fraser and Wilson and Johan exercised agency in planning for fatherhood/parenthood by deliberately seeking out lesbian couples with whom they could form co-parenting arrangements. In the third couple, Max, who donated for a lesbian couple at their request, also exercised agency in becoming a father/parent. Controlling the circumstances in which this came about, Max expected the women to respond positively to his long-term view of himself as a father/parent as a condition of donating, despite that not having been their original plan. His (then) new partner Patrick gradually adopted the same long-term view. While known donor-child relationships and roles may be mediated and controlled by mothers, who typically determine the extent to which relationships and roles develop (Hertz, 2002), the picture appears more complex for each of the six men in this chapter. The men's exercise of agency extends to the strategic adoption or acceptance of a subordinate status relative to the lesbian couples they imagine or actually collaborate with, because this works for them.

Wilson and Johan and Max and Patrick draw on a discourse of paternal choice and established patterns for non-resident fathers/parents in the divorce or separation context as key resources shaping their stories. Vivian and Moira and Nicole and Jeannie assign particular father-child relationships and roles to them, including non-resident father/parent with part time participation in care and decision-making practices. Although this could be read as a strategy by female parents to bring the men under the women's direction, the men embrace their subordinate status. The women's residential primary parenthood is useful to the men because it allows them to achieve their preferred form of fatherhood/parenthood. In this *remarkable situation*, they *get to live that life… to be parents*, by choosing parenting as a supplement to their social and working lives, rather than as a central focus. Because the men intend to or actually outsource childcare to the women the majority of the time, they can experience the rewards of parenting without the mundane daily grind mothers often experience. In this way, they are implicated in processes of normalisation that duplicate traditional gendered parenting in heterosexual households, while also engaging in innovative cross-household multi-parenting.

Kole and Fraser's preference for a form of fatherhood/parenthood inclusive of this daily grind rejects a discourse of paternal choice and divorced *weekend dads* as resources in their stories. In their case, a lesbian couple willing to share parenting equitably with them will allow them to fulfil this aspiration. Regardless, they are prepared to embrace a subordinate status relative to a lesbian couple, because a lesbian couple will also be useful to them as female parents. Wishing to safeguard their children from the *big issue* of *the missing parent*, female parents will function as protective factors in their future children's lives, sheltering them against loss and harm. They use their narrative to construct certain sorts of personally responsible parenting selves and identities, by taking their future children's needs into account and weighing up the possible consequences of their choices. Rather than internalising norms that suggest children will suffer if they do not know their father, Kole and Fraser have internalised parallel norms relative to mothers. These norms, which include the heteronormative expectation of opposite sex parents, are important resources for them.

The couples' narratives perform identity work. They illustrate what kinds of fathers/parents the couples imagine they will become or believe themselves to be and what matters to them at particular moments in time. Their stories construct them as new fathers who are the ideological equivalent of the mothers, even where they are not expected to be or actually are their equivalent in practical terms.

The couples reconfigure notions of what it means to be a father/parent in several ways. All three couples separate biogenetic fatherhood/parenthood from the doing of fathering/parenting through reflexive negotiation of expected or actual levels of involvement with children, with one couple disrupting the assumption that a female body is a prerequisite for mothering and dominant hegemonic masculinities that position fathers/parents in traditional ways. As Stacey (2006) found, 'Gay fatherhood… represents terrain more akin to motherhood than to dominant forms of heterosexual paternity' (p. 48) (see also, Berkowitz, 2011; Ó Súilleabháin, 2017; Pannozzo, 2014). Finally, the couples disconnect fathering, mothering, and joint residence. 'Home', for their planned and actual children, is a shifting space, attached to relationships, rather than places (Donovan, 2000).

The men's recourse to the narrative resources mentioned in this discussion supports the identity work the stories accomplish. While they exercise agency and choice about how to be a father/parent and strategically adopt or accept a subordinate status relative to the lesbian couples, their fatherhood/parenthood is simultaneously expanded and curtailed by these resources. The men reconfigure notions of what it means to be a father/parent, even as they reinforce traditional meanings, with this reinforcement lending their subordinate status an inevitable quality. In sum, they narrate their self-as-father and self-as-parent identities through an existing mix of heteronormative tropes; while their narratives reflect the dominant social order, their self and identity construction work does challenge this order at times.

Notes

1. In Aotearoa, two options for legal parenthood are available in cases like theirs; either the biogenetic mother and father can become legal parents, or the biogenetic mother and her female partner can become legal parents through the provisions of the Status of Children Amendment Act 2004, Part 2.
2. Johan was not available to be interviewed.

References

Andreasson, J., & Johansson, T. (2017). It all starts now! Gay men and fatherhood in Sweden. *Journal of GLBT Family Studies, 13*(5), 478–497. https://doi.org/10.1080/1550428X.2017.1308847

Berkowitz, D. (2011). Maternal instincts, biological clocks, and soccer moms: Gay men's parenting and family narratives. *Symbolic Interaction, 34*(4), 514–535. https://doi.org/10.1525/si.2011.34.4.514

Bold, C. (2012). *Using narrative in research*. Sage Publications Ltd. http://dx.doi.org.ezproxy.canterbury.ac.nz/10.4135/9781446288160

Brockmeier, J., & Carbaugh, D. (2001). Introduction. In J. Brockmeier & D. Carbaugh (Eds.), *Narrative and identity: Studies in autobiography, self and culture* (Vol. 1, pp. 1–22). John Benjamins Publishing Company.

Carneiro, F. A., Tasker, F., Salinas-Quiroz, F., Leal, I., & Costa, P. A. (2017). Are the fathers alright? A systematic and critical review of studies on gay and bisexual fatherhood. *Frontiers in Psychology, 8*. https://doi.org/10.3389/fpsyg.2017.01636

Carsten, J. (2004). *After kinship*. Cambridge University Press.

Chisholm, D. (2016). Gene pull. *New Zealand Listener, 254*(3968), 14–21.

Côté, I., & Lavoie, K. (2019). A child wanted by two, conceived by several: Lesbian-parent families negotiating procreation with a known donor. *Journal of GLBT Family Studies, 15*(2), 165–185. https://doi.org/10.1080/1550428X.2018.1459216

Dempsey, D. (2004). Donor, father or parent? Conceiving paternity in the Australian Family Court. *International Journal of Law, Policy and the Family, 18*(1), 76–102.

Dempsey, D. (2012a). Gay male couples' paternal involvement in lesbian-parented families. *Journal of Family Studies, 18*(2/3), 155–164. https://doi.org/10.5172/jfs.2012.18.2-3.155

Dempsey, D. (2012b). More like a donor or more like a father? Gay men's concepts of relatedness to children. *Sexualities, 15*(2), 156–174.

Donovan, C. (2000). Who needs a father? Negotiating biological fatherhood in British lesbian families using self-insemination. *Sexualities, 3*(2), 149–164. https://doi.org/10.1177/136346000003002003

Dunne, G. A. (2000). Opting into motherhood: Lesbians blurring the boundaries and transforming the meaning of parenthood and kinship. *Gender & Society, 14*(1), 11–35.

Fertility Associates. (n.d.). Waitlist for donor sperm. https://www.fertility associates.co.nz/treatment-options/donor-options-and-surrogacy/donor-sperm-waitlist/

Finch, J., & Mason, J. (2000). *Passing on: Kinship and inheritance in England*. Routledge.

Griffin, G. (2020). "It's just sperm. That's all your're giving": Men's views of sperm donation. In G. Griffin & D. Leibetseder (Eds.), *Bodily interventions and intimate labour: Understanding bioprecarity* (pp. 131–146). https://www.jstor.org/stable/j.ctvwh8fh8.12

Hayman, B., Wilkes, L., Halcomb, E., & Jackson, D. (2014). Lesbian women choosing motherhood: The journey to conception. *Journal of GLBT Family Studies, 11*(4), 395–409. https://doi.org/10.1080/1550428X.2014.921801

Herbrand, C. (2017). Co-parenting arrangements in lesbian and gay families: When the 'mum and dad' ideal generates innovative family forms. *Families, Relationships and Societies, 7*(3), 449–466. https://doi.org/10.1332/204674317X14888886530269

Hertz, R. (2002). The father as an idea: A challenge to kinship boundaries by single mothers. *Symbolic Interaction, 25*(1), 1–31. https://doi.org/10.1525/si.2002.25.1.1

Kelly, F. J. (2011). *Transforming law's family: The legal recognition of planned lesbian motherhood*. The University of British Columbia Press.

Mallon, G. P. (2004). *Gay men choosing parenthood*. Columbia University Press.

Mason, J. (2004). Personal narratives, relational selves: Residential histories in the living and telling. *The Sociological Review, 52*(2), 162–179. https://doi.org/10.1111/j.1467-954X.2004.00463.x

Mason, J. (2011). What it means to be related. In V. May (Ed.), *Sociology of personal life* (pp. 59–71). Palgrave Macmillan.

Nebeling Petersen, M. (2018). Becoming gay fathers through transnational commercial surrogacy. *Journal of Family Issues, 39*(3), 693–719. https://doi.org/10.1177/0192513X16676859

Nordqvist, P. (2012). Origin and originators: Lesbian couples negotiating parental identities and sperm donor conception. *Culture, Health & Sexuality: An International Journal for Research, Intervention and Care, 14*(3), 297–311. https://doi.org/10.1080/13691058.2011.639392

Nordqvist, P. (2019a). Kinship: How being related matters in personal life. In V. May & P. Nordqvist (Eds.), *Sociology of personal life* (2nd ed., pp. 46–59). Macmillan International.

Nordqvist, P. (2019b). Un/familiar connections: On the relevance of a sociology of personal life for exploring egg and sperm donation. *Sociology of Health & Illness, 41*(3), 601–615. https://doi.org/10.1111/1467-9566.12862

Nordqvist, P. (2020). Bioprecarity and pregnancy in lesbian kinship. In G. Griffin & D. Leibetseder (Eds.), *Bodily interventions and intimate labour: Understanding bioprecarity* (pp. 95–110). https://www.jstor.org/stable/j.ctvwh8fh8.10

Ó Súilleabháin, F. (2017). Expanding 'Irish family' repertoires: Exploring gay men's experiences as parents in the Republic of Ireland. *Journal of GLBT Family Studies, 13*(5), 498–515. https://doi.org/10.1080/1550428X.2017.1308848

Pannozzo, D. (2014). Child care responsibility in gay male-parented families: Predictive and correlative factors. *Journal of GLBT Family Studies, 11*(3), 248–277. https://doi.org/10.1080/1550428X.2014.947461

Park, N., Kazyak, E., & Slauson-Blevins, K. (2015). How law shapes experiences of parenthood for same-sex couples. *Journal of GLBT Family Studies, 12*(2), 115–137. https://doi.org/10.1080/1550428X.2015.1011818

Riessman, C. K. (2008). *Narrative methods for the human sciences*. Sage Publications, Inc.

Riggs, D. W. (2008a). Lesbian mothers, gay sperm donors, and community: Ensuring the well-being of children and families. *Health Sociology Review, 17*(3), 226–234.

Riggs, D. W. (2008b). Using multinomial logistic regression analysis to develop a model of Australian gay and heterosexual sperm donors' motivations and beliefs. *International Journal of Emerging Technologies and Society, 6*(2), 106–123.

Smart, C. (2007). *Personal life: New directions in sociological thinking*. Polity Press.

Smart, C. (2011). Relationality and socio-cultural theories of family life. In R. Jallinoja & E. D. Widmer (Eds.), *Families and kinship in contemporary Europe. Rules and practices of relatedness* (pp. 13–28). Palgrave Macmillan Ltd.

Sparkes, A. C. (2005). Narrative analysis: Exploring the whats and hows of personal stories. In I. Holloway (Ed.), *Qualitative research in health care* (pp. 191–209). Open University Press.

Stacey, J. (2006). Gay parenthood and the decline of paternity as we knew it. *Sexualities, 9*(1), 27–55.

Stevens, E. (2015). Understanding discursive barriers to involved fatherhood: The case of Australian stay-at-home fathers. *Journal of Family Studies, 21*(1), 22–37. https://doi.org/10.1080/13229400.2015.1020989

Surtees, N. (2011). Family law in New Zealand: The benefits and costs for gay men, lesbians, and their children. *Journal of GLBT Family Studies, 7*(3), 245–263.

Surtees, N. (2017). *Narrating connections and boundaries: Constructing relatedness in lesbian known donor familial configurations* (PhD thesis). University of Canterbury, New Zealand.

Surtees, N., & Bremner, P. (2020). Gay and lesbian collaborative co-parenting in New Zealand and the United Kingdom: 'The law doesn't protect the third parent'. *Social & Legal Studies, 29*(4), 507–526. https://doi.org/10.1177/0964663919874861

12

Doing Reflexivity in Research on Donor Conception: Examining Moments of Bonding and Becoming

Giselle Newton

Introduction

Researchers often hold significant personal investments in their fields of study (Taylor, 2011). The reflexive turn in the social sciences has led to greater transparency around knowledge construction processes (Mauthner & Doucet, 2003; 2018). By practicing reflexivity, that is, critical reflection in relation to knowledge construction, researchers can openly monitor the ways in which their subjectivity affects all aspects of the research process (Nathan et al., 2018). As Guillemin and Gillam (2004) argue, in addition to contributing to rigorous research practice, reflexivity forms part of 'ethics in practice' (p. 262). In this way, researchers are sensitised to ethical and interpersonal issues that emerge in research on an everyday basis. In donor conception studies, limited attention has been paid to the way researchers situate themselves in relation to research participants or ethical issues that emerge through the research

G. Newton (✉)
Centre for Social Research in Health, UNSW, Sydney, NSW, Australia
e-mail: g.newton@unsw.edu.au

© The Author(s), under exclusive license to Springer Nature Singapore Pte Ltd. 2022
R. M. Shaw (ed.), *Reproductive Citizenship*, Health, Technology and Society,
https://doi.org/10.1007/978-981-16-9451-6_12

process; a remarkable oversight given the emotionally and politically sensitive nature of this topic (for an exception see Kirkman & Kirkman, 2002). Yet qualitative researchers in particular may have a unique opportunity to capture the tensions that arise in research on donor conception, particularly in relation to positionality and research practice (see also Shaw et al., 2020). In this chapter, I explore what it means to 'do reflexivity' in my research on donor conception, illuminating a range of ethical and emotional issues that have arisen for me as an insider researcher.

Commencing a PhD focused on donor conception prompted my own personal reflections on my experiences as a donor-conceived person and motivated me to think carefully about my positionality. 'Positionality' refers to all of my personal characteristics that influence my access to the field, relationship to participants and research process (Berger, 2015). Some scholars have critiqued static observations of positionality as a "laundry list of identity markers" (Kohl & McCutcheon, 2015, p. 747), and consequently, several frameworks have been developed for 'doing reflexivity'. Of these frameworks, Guillemin and Gillam's (2004, p. 262) focus on 'ethical moments', defined as 'the difficult, often subtle, and usually unpredictable situations that arise in the practice of doing research', has offered a tangible method for reflecting on the research process, emotional connections to research and the role of the researcher. Another framework by Fox and Allan (2014) consists of dialogue between the doctoral student researcher and their supervisor. This approach aims to promote 'reflexive action' (p. 101) by recalling salient ethical, performative and conceptual moments during the student's studies that have impacted the student's identity and influenced the course of the research. The authors describe these instances as moments of becoming and unbecoming, recognising the unravelling and re-formation that occurs both to the research and, consequently, to the researcher's identity. These methods for 'doing reflexivity' demonstrate that reflexivity is not static nor individual, but rather an ongoing negotiation between actors involved in the research process.

Based at UNSW Sydney, Australia, my research examines Australian donor-conceived people's experiences, perspectives and support needs. I consider how donor-conceived people's sense of identity, community and belonging is forged across different contexts. For example, I explore what donor conception means to participants at an individual level, within their own families, as well as what it feels like to belong to a community of donor-conceived people. Specifically, I consider how participation in secret Facebook groups, direct-to-consumer DNA testing platforms or national inquiries impact participants' experiences. This study integrates sociological, social semiotic and media studies perspectives, considering participants' interpretations of their experiences as well as analysing the often less conscious language choices they make. The choice to integrate these distinct approaches was made due to the diverse and complex nature of donor-conceived people's lived experience. As Mason (2006) argues, our social lives and realities are multidimensional, and mixing methods can help to simultaneously transcend both micro and macro domains of lived experience.

In attempting to synthesise these disciplinary methodologies, significant ontological and epistemological challenges arise. For example, social semiotics is concerned with understanding language in context, collecting data from language use in situational contexts (Halliday, 1978). In sociology, the co-construction of the knowledge produced through the research encounter is foregrounded (Braun & Clarke, 2013). I seek to integrate sociological approaches to belonging with a social semiotic approach that views language as a social resource through and within which relationships are enacted.

In this chapter, I describe three consecutive overlapping phases: becoming donor-conceived, becoming activist and becoming researcher. My experience of 'becoming researcher' is deeply entangled with my position as a donor-conceived person and activist, as the subjectivities I simultaneously hold are not autonomous; they are always inflected by the other positions. This perspective aligns with an intersectional approach from scholars such as Crenshaw (1991), Collins (1998) and Few-Demo (2014), who recognise that social identities are constituted by multiple social locations that may overlap in different ways across time and space. Considering my research through the lens of each of these positionalities, and the resulting tensions between them, provides insight into the particularities of conducting research on donor conception as a donor-conceived person myself. In reflecting on my personal experiences, I hope to stimulate discussion in donor conception studies about how we can continue to do reflexivity in this field of research.

Becoming Donor-Conceived

Throughout its history, the practice of donor conception has often been clouded in secrecy, shame and denial, reflecting the complex social, religious and legal contexts that govern reproduction and gamete donation more generally (Crawshaw, 2018; Daniels, 2020; Frith et al., 2018). In Australia, while legislation varies between states, historically all states have favoured the rights of donors to lifelong anonymity over the rights

> *I have a vague memory that I've played on repeat thousands of times throughout my life, where my mother and father are sitting at our round dining room table with a stone blue vinyl tablecloth. I think I was 8 or maybe 10. They have something they have to tell me, and I run away, down the corridor to my bedroom. When I think about this now, I don't really know why I ran away—they'd been preparing me for this conversation my whole life. To their credit, despite the shame that infertility can carry, and their doctor's advice to deceive me, my parents always weaved the narrative about my donor conception into my sense of identity. They told me how difficult it was for them to have a baby and that they were so lucky to have me. That day that I ran away, we ended up having a conversation about the details of my conception, but it didn't come as a shock to me that I was donor-conceived.*
>
> *Growing up in a regional town, I didn't know any donor-conceived people, didn't really know what being donor-conceived meant, and didn't think about it too much. Around the time I turned eighteen, I joined a secret Facebook group for Australian donor-conceived people. I began to see donor-conceived people's stories on a daily basis. Through this contact with other people like me, I began learning about what being donor-conceived meant.*
>
> *I write my story into my research because it is embedded in a history of secrecy. I believe it is necessary to recognise my position as a donor-conceived peer—the degree to which this situatedness influences my research approach means that obscuring it would be unethical. I now understand that I am part of a network of shared experiences and values, and that my personal biography is part of a broader social and political context that extends beyond my individual and family relationships.*

of donor-conceived people to access identifying information about their donor. Driven by increased social and legal recognition of the rights of the child, law reform over the last two decades has largely sought to improve the rights of donor-conceived people to access information, and to restrict the rights of donors to remain anonymous. While disclosure of donor-conceived status to children is now widely recognised as best practice (Frith et al., 2018), generations of parents of donor-conceived people were advised by clinicians not to tell their children of their donor-conceived status, and to not seek out identifying information about their donors (Adams & Lorbach, 2012). Since the conception of Louise Brown, the first person conceived via in vitro fertilisation in 1978, the fertility industry has grown rapidly. In 2018, over 19,000 babies were born using ART in Australia and New Zealand, including 406 babies through donor sperm insemination, 761 babies from oöcyte or embryo donation and 86 through gestational surrogacies (Newman et al., 2020). While there is no way to definitively determine the number of donor-conceived people in Australia, estimates suggest that there may be between 20,000 and 60,000 persons (Commonwealth of Australia, 2011). Yet, despite the significant number of people born through donor conception, donor-conceived people have been described as a 'hidden population' (Hertz & Nelson, 2018).

The subject of donor conception has attracted a remarkably limited range of research. Prior to 2000, no empirical studies had been published on donor-conceived people or parents' experiences of donor conception and from 2000 to 2011, 19 articles derived from empirical research on donor conception were published (Blyth et al., 2012). Studies employing quantitative methods to explore the experiences of donor families comprise most of the existing literature (Andreassen, 2017), and a large portion of that research has sourced participants from the US-based donor registry, Donor Sibling Register (see, e.g., Jadva et al., 2010) or UK-based DNA Link registry (see, e.g., Crawshaw & Marshall, 2008).

Existing research has found that the majority of donor-conceived people view any information about their donors as important (Rodino et al., 2011), and most desire contact with their donor (Beeson et al., 2011) and/or donor siblings (Nelson et al., 2013), including interpersonal contact (Dempsey et al., 2019). For those who do achieve contact, most consider it to have been a positive experience (Jadva et al., 2010). It is also worth noting that, since the topic has received more scholarly attention, research has largely focused on parents' attitudes towards the disclosure of donor-conceived status and towards connecting with other families who have used the same donor. However, parental views represent only one set of perspectives on donor conception, and these perspectives are likely to have been shaped by social desirability bias, given the parental desire to believe their children are happy, healthy and well supported, no matter the choices their parents made about their conception (see Macmillan, 2016).

Even fewer studies have focused on the experience of Australian donor-conceived people. In 2002, Australian donor-conceived activist Geraldine Hewitt focused her final year high school project on the identity issues experienced by 47 donor-conceived people. In her concluding remarks, she stated that: 'it is poignant that this study is, to date, the largest international study of the individuals who have been conceived through donor insemination' (p. 5). Hewitt's (2002) informal study, conducted in a high school setting, has received a remarkable number of citations, indicating that there is an appetite for donor-conceived voices to be heard and understood. The same year, Eric Blyth (2002b) noted that, to his knowledge, the few studies of donor-conceived adults' experiences had all been conducted by donor-conceived people. Little has changed since 2002, with donor-conceived scholars still producing much of the research on donor-conceived people's experiences and perspectives. Donor-conceived scholars have outlined the need for a child-centred perspective (Adams, 2013; Rose, 2009) and discussed the impact of secrecy on their lives (Whipp, 1998). In a similar vein, Macmillan's (2016) study of 69 Australian donor-conceived adults was the first to consider the psychosocial implications of donor conception, secrecy and anonymity for donor-conceived adults. More recently, Martin Eggen Mogseth (2019) explored the impacts of donor conception on identity and familial relationships, and Damian Adams' (2021) study was the first to explore the health outcomes of adult donor-conceived people. More recently, other scholars such as Ken Daniels (2020) have recognised the importance of capturing adult donor-conceived people's experiences and views and noted the unique insights donor-conceived people bring to examining past and present practices, policies and attitudes.

Returning to the vignette above, and the theme of 'becoming donor-conceived', I am often asked how I found out I was donor-conceived. However, perhaps a more influential experience, in terms of becoming and belonging, was joining a Facebook group for donor-conceived people and gaining access to a community who shared that aspect of my identity. Among donor-conceived people, there is a strong history of peer-led organisations, as well as online groups, established as safe spaces to discuss experiences and connect with those who may feel isolated from a lack of contact with donor-conceived peers (Adams & Lorbach, 2012). Having peers to connect with and relate to strongly influenced the way that I understood my own experience of being donor-conceived. This has shaped my belief that we cannot understand donor-conceived identities without also seeking to understand how meaning emerges through donor-conceived communities, recognising that identity is negotiated in interaction with communities of shared values (Knight, 2010).

Since the turn of the century, the internet has had a significant impact on the way donor-conceived people access information about their donor, their donor siblings and other aspects of their conception story (Blyth, 2002a). Internationally, donor-conceived people have been supporting each other since the advent of the Yahoo group 'People Conceived Via Artificial Insemination' in 2000. In Australia, the Australian Donor Conception Forum was founded in 2006 (Adams & Lorbach, 2012). Today, social media groups are used by donor-conceived people to access information, seek advice and share their experiences. Personally, being able to access a community of people 'like me' has opened up new avenues for me to seek information about my conception story, the meanings I have ascribed to being donor-conceived and the extent and frequency of contact I have with other donor-conceived people. For example, like many donor-conceived people, I would not have thought to explore DNA testing platforms as a mechanism for initiating contact with my genetic family. However, after hearing repeated stories of donor-conceived people finding their donor or donor siblings via DNA matching, I felt that this could be a significant opportunity to gain information about my own conception history, especially after tiresome battles with hospitals, clinics and registers to seek information about my donor. I did not know that 1737 centimorgans equated to a sibling match, but I learned that on a Facebook group. When I matched with my donor sibling, who may not have known that they were donor-conceived, I did not know how to approach them, and so it was incredibly helpful to be able to access advice and guidance from peers who had already managed these challenges.

I continue to be surprised and moved by the considerable time donor-conceived people take to show empathy and kindness to strangers online. I was curious as to what motivated them to provide lengthy, detailed and supportive comments in Facebook groups for donor-conceived people. From a linguistic perspective, I interpret this as a form of affiliation: a model which views social bonds as the social semiotic unit for building identities and communities (Knight, 2010). Eggins and Slade (1997) coined the term 'orientation to affiliation' to describe 'the extent to which we seek to identify with the values and beliefs of those we interact with, especially in perceiving others as insiders or outsiders' (p. 53). I seek to understand how donor-conceived people negotiate attitudes, and bond around shared values, a social process that contributes to a sense of belonging. While online support groups for donor-conceived people have received little scholarly attention, my personal experience has stimulated my interest in the way online groups influence donor-conceived people's sense of belonging. Indeed, in addition to supporting my process of understanding my own story, these online communities also accelerated another transition for me, this time to 'becoming activist'.

Becoming Activist

The phrase 'nothing about us without us' (Charlton, 2000), coined in the early 1990s by those in the Disability Rights movement, encapsulates the importance of engaging people with lived experience in decision-making about programme and service development, as well as leadership and governance. In the health sector, people with lived experience are increasingly being recognised as experts on issues affecting their own health in the form of 'consumer participation', defined as 'incorporating consumer opinions

I have never been among so many donor-conceived people in real life. We are packed into a tiny room with barely enough space to get out of our chairs in an office building in the city centre of Geneva. Everyone is taking turns to read through the speech they will present at the UN the following day. Many members of the group have spent decades pushing for reform. The room is hot, filled with nerves in preparation for this milestone event. The organisers, Belgian donor-conceived activist Stephanie Raeymaekers, and Australian health law expert and advocate Professor Sonia Allan, have put hours of work into organising this workshop at the 30th Anniversary of the UN Convention on the Rights of the Child (UNCRC). We have paid our way, travelling from all corners of the world to discuss our experiences of donor conception on the world stage for the first time (Allan et al., 2020).

Those who are delivering speeches at the UN the next day are tasked with telling emotive and persuasive personal stories, rehashing sensitive and painful experiences with the aim of influencing decision-makers. As people rehearse their stories, highlighting the consequences of donor conception, tissues are handed around as they become emotional. Some tell stories of finding out they have dozens of siblings, others of searching for information about biological parents for decades. The most emotional of the stories are of donor-conceived friends, such as Narelle Grech and Alison Davenport, who tragically passed away from conditions for which they were high-risk, information they would have known had they been permitted access to the medical histories of their donors. Legislation across the globe has failed donor-conceived people by neglecting to recognise our right to know our identities and our families. The speeches are connected by the shared vision of "nothing about us, without us"; governments must listen to and act on the voices of people with lived experience. The next day, the presentation is met with a standing ovation, members of the UN and international community moved to tears.

My trip to Geneva emphasised to me that our stories have political value, and that our voices must be given primacy. I became more aware of how I could leverage my experience to help change laws and support the rights of generations of donor-conceived people. I also witnessed how donor-conceived people bonded around shared adversity and the struggle to have their human rights recognised, and participated in the celebration of feeling listened to and creating change. Reflecting on that moment, it is clear to me how activist involvement shifted my priorities and my approach to research. I now believe that researchers in the field of donor conception have a key role to play as allies and advocates, and that they possess the power to centre the narratives of donor-conceived people. Ignoring donor-conceived people's voices and priorities is untenable.

and perspectives to inform the development and improvement of healthcare practice' (Hall et al., 2018, 708). The beginning of the consumer participation movement was sparked by the reform of the mental health care system in the US in the 1970s (Doyle, 2008), and while the appropriateness of using the term 'consumer' in health contexts has been debated (Smith, 2016), the importance of this mode of participation and representation is now widely recognised (World Health Organization, 1978). The large body of scholarship on consumer participation has demonstrated that governments and organisations that act without collaborating with people with lived experience will inevitably have less significance and limited outcomes. For example, Ti et al. (2012) have argued that collaboration between decision makers, service providers and people with lived experience strengthens community voices and results in greater satisfaction and wellbeing. However, in practice, the ability to uphold these standards proves challenging for service providers, as it requires systematic change to include people with lived experience in meaningful ways.

ART brings with it considerable scientific and medical complexities, as well as significant social and ethical issues. The social and ethical issues largely stem from balancing the interests of a wide range of stakeholders, including those of intending or recipient parents, clinicians and allied health professionals, fertility clinics and their shareholders, regulators and taxpayers. However, people conceived through ART are seldom counted among this set of key stakeholders, despite models encouraging a child-centric paradigm (Adams, 2013). Donor-conceived people frequently note that they are the only party who do not consent to this conception arrangement, while being the party most affected by the decision. In the opening of her doctoral thesis, donor-conceived woman Joanna Rose suggested that 'encouraging social change which goes against powerful interests, is invariably difficult and painful ... To my delight, I have found there were people fighting, against the odds, for our rights from before I was born' (Rose, 2009, p. viii). Standing against a multi-billion-dollar industry takes significant emotional energy. However, given there are now several generations of people who know they were donor-conceived:

> It is paradoxical that those for whom the practice was created in the first place have not been an integral part of academic and practical discussions

on secrecy, anonymity, disclosure, and other aspects of its practice that affect their lives in fundamental ways. (Mahlstedt et al., 2010, p. 2236)

Among the literature on donor conception, it is widely recognised that the best strategy for achieving the 'best interests' and 'welfare' of donor-conceived people is to ask them (Blyth, 1998). As suggested by Rodino et al. (2011), donor-conceived people 'may be best placed to guide policy makers as to the type of information that should be stored and made available because they have been most affected by the process' (p. 310). However, limited attention has been paid to the ways in which donor-conceived people are permitted to participate in shaping policy and law regarding assisted reproductive technology based on their lived experience. This raises two questions: are donor-conceived people being included in decision-making, and, if they are, what effect does their participation have on policy and practice?

While few studies have focused on donor-conceived people's perspectives, it has been noted that the voices of donor-conceived adults are 'very strong, influential and demanding to be heard' (Daniels, 2020), and that donor-conceived people are extremely willing to be involved in research (Macmillan, 2016). The feelings of powerlessness which donor-conceived people may experience in relation to their genetic histories may be reinforced when they are excluded from research that affects them. Meaningful inclusion of people with lived experience in research can take many forms, from participatory research (such as Fisher & Robinson, 2010) to advisory committees (see Porter et al., 2006). Scholars such as Schneider (2012) have noted that there is a continuum of approaches to inclusion, from control, collaboration, contribution and consultation (p. 156). In the field of donor conception studies, there is need for a greater focus on what it means to do research with and for donor-conceived people.

In my research, I decided to privilege donor-conceived people's perspectives by focusing exclusively on their experiences rather than 'balancing' their voices with those of donors, recipient parents and clinicians. I was also interested in donor-conceived people's opinions on the practice of donor conception itself, to provide context for what it is they need from and bring to online support communities. Consequently, I sought to gather donor-conceived people's views about payment or reimbursement for gametes, limits on the number of children born from the same donor, disclosure about donor conception (including on birth certificates) and importation or travel to access gametes. In doing so, I sought to deliberately recognise donor-conceived people as experts on the issue. I also wanted to examine whether donor-conceived people felt they were being included in decision-making and development of programmes and services, and whether they had been involved in their design and delivery. Finally, I opted to include donor-conceived people in the governance of the research. In the design phase of my study, I established an advisory panel to provide expert advice on the study, which included people who have lived experience of being donor conceived in an Australian context. Involving donor-conceived people in a research approach promotes community ownership and facilitates recruitment, which is especially valuable for empirical study with 'hard-to-reach' groups.

The decision to focus on donor-conceived people's perspectives is itself political. The right to know one's genetic origins is a human right (UNICEF, 1989), and this human rights framework is undisputed within my community. However, these perspectives can come into conflict with other rights claims, and there are people in the community who may disagree with donor-conceived people being able to access identifying information. I am acutely aware that, because of my dual positions as donor-conceived person and as activist, I may not be viewed as an objective researcher, and my research may come under additional scrutiny. Every researcher brings their own insights and biases to their research, although for peer researchers those insights and biases are often subject to greater critique. Bringing my lived experience to my research gives it ethical legitimacy, but also opens the research and myself up to opposition. Flood et al. (2013) have argued that activist researchers must protect themselves, firstly from opposition from political opponents who may aim to silence them, as well as from their institutions who may view activism as going beyond the researcher role. I am particularly cognisant of these tensions as I move into my third position of 'becoming researcher'.

Becoming Researcher

> I'm with a friend on a Friday night in the middle of winter. I open Facebook to have a quick scroll. A post on my feed catches my attention. It's from one of the Facebook groups for donor-conceived people, describing how the person had just found out they were donor-conceived, as an adult, in the middle of the coronavirus pandemic. I drop a care reaction, feeling for this person who is newly discovering and trying to make sense of this information, and move back to my conversation with my friend. The following week, the person from the Facebook post reaches out to me to do an interview for my study. I recognise the name, and I wonder if it's too soon to be doing an interview with them, and whether the interview may be distressing. I decide to proceed with the interview because participants often tell me that they find participating in the research therapeutic, an opportunity to chat with someone who understands.
>
> As I am scrolling through Facebook, I often stumble upon snippets of information relevant to my research, meaning that I may know things about a participant before we meet. I feel unsettled knowing part of a participant's story before the interview encounter, because they have disclosed that information to me as a peer, not as a researcher.
>
> Despite my discomfort, by agreeing to do the interview, I was also able to check that they were doing okay and provide them with resources. Following the interview, the participant told me that they enjoyed talking to a fellow donor-conceived person and asked to add me on Facebook to keep in touch. This demonstrates several ethical tensions that have emerged from my new and overlapping positions as peer and researcher. For researchers who investigate their own community, the boundaries between work hours and home hours, research and personal life, can become blurred.

My positionality as a donor-conceived person means that I come at my research as an insider. There is a large body of scholarship in qualitative research focused on positionality in relation to research participants which characterises possible impacts of a researcher's insider or outsider position, although much of what has been written comes from the areas of field research, ethnography and observation (Dwyer & Buckle, 2009). As suggested by Dwyer and Buckle (2009), being an insider does not inherently make the research better or worse; rather, coming at a research topic from different angles can enrich our understanding. In particular, insider researcher can affect the rapport between the researcher and participants as 'shared experiences cultivate degrees of intimacy between people' (Taylor, 2011, p. 10). Some scholars have critiqued the term 'insider' as 'dichotomised rubrics such as "black/white" or "insider/outsider" are inadequate to capture the complex and multi-faceted experiences of some researchers' (Song & Parker, 1995, p. 243). Further, subjectivities are not static: in interaction, different aspects of identity become more or less prominent depending on audience and context, thus one is neither entirely insider nor entirely outsider (Hodkinson, 2005). Hodkinson (2005) argues that in particular situations, despite identity differences, participants are consciously united by a key feature or set of features. The identity category 'donor-conceived' is one that those conceived via third-party reproduction may or may not identify with, and among those who do identify as donor-conceived, individuals may inhabit a range of social locations. Therefore, remaining cognisant of the divide between the stories of the participants and my own experiences of donor conception is vital. For instance, in insider research with fat women, Howe outlines the importance of not overshadowing participants' perspectives and experiences with her own stories or generalising the experiences of all fat women (Shaw et al., 2020). Berger (2015) has described this as a 'constant deliberate effort to maintain the separation between mine and theirs' (p. 224).

When I started my research, I was nervous about how it might be perceived by my community, and wary of negative perceptions that could impact my personal life. Research by Johnson et al.. (2018), on online communities for military spouses, explored the risks involved in doing research on one's own online community, particularly emotional risks, such as social isolation, if the research or researcher is not well perceived by the community. However, my major concerns did not come to fruition, rather the affinities created in the interview context have enhanced my sense of belonging to a community of donor-conceived people. As I was conducting interviews, many participants told me that they trusted me to do a good job with the research because I was a peer, that they were able to better open up as a consequence, or that I 'got it'. As discussed by Attia and Edge (2017), for many insider researchers, empirical work is built upon pre-existing trust. Then, through the research process, interactions with community members increase, and relationships are strengthened. Increased contact with my peers helped me to become more confident in my identity as a donor-conceived person, and, in turn, I now feel more a part of the Australian donor conception community as I am recognised by them as a peer.

My positions as a researcher and donor-conceived person often collide when the expectations of each role conflict. To make sense of these competing norms between my roles, I use the different disciplinary methodologies I have access to. When I consider the tension between peer and researcher from a linguistic perspective, for example, I first consider how social context influences my interviewer role and the interaction patterns I can produce. In social semiotics, types of interaction are understood as genres; every type of interaction has stages, goals and social factors (Martin, 2009). In the genre of a qualitative interview, the goal is to ask questions and elicit information from the participant without leading the participant to give a biased or socially desirable response. Therefore, to conform to the requirements of the researcher role, I consciously employ strategies that minimise the extent to which I align or de-align with the values expressed by the participant. At an interpersonal level, maintaining neutrality is challenging, because as a donor-conceived person I feel compelled to align with the participant as a peer in our interaction.

There are several strategies, deemed appropriate within qualitative research, that I use to build connection and rapport at the beginning and end of interviews. By flagging that I am donor-conceived at the start of interviews and demonstrating knowledge of appropriate word choice and understanding of what is going on in the community, I build trust with participants. Further, I conclude interviews by asking participants if they have any questions for me and I find that participants often use this opportunity to ask about my story and to find ways to connect our experiences. In such instances, I think it is appropriate to draw on the similarities between my own experience and the participant's in seeking to validate their interpretation. I also use this opportunity to recognise the contributions they have made to the research by sharing their stories with me. Using the final moments of the interview to do this allows me to take off my 'researcher hat' and connect peer to peer, while also ensuring that participants feel appreciated. Julie Howe has described the experience of a participant reaching out to hug her following an interview and the resulting tension between doing what is 'right' as a kind human being and 'proper research conventions' (Shaw et al., 2020, p. 282). Like Howe, I believe that through building reciprocal relationships with participants, we strengthen the quality of our data and permit an emotional connection.

The literature on insider research has often focused on participants' wellbeing, and the emotional toll that insider research can have on the researcher has been explored less (see Nelson, 2020). During this research I have been mindful of assessing how I am feeling given that the topic is so personal. While it is recognised that interviews can be cathartic or therapeutic for participants (Shaw et al., 2020), less has been written about how the researcher may also benefit from the interaction. I have found meeting donor-conceived people and hearing their stories to be affirming. Reflecting on their data collection, Nelson (2020) has suggested that:

> Through hearing the diversity of stories and experiences of people who were like me/not like me, I saw a variety of opportunities, experiences and feelings that I had never had the time to consider in such depth before. (p. 13)

Like Nelson, I feel a very deep and personal connection to my interview participants, despite the diverse range of stories and experiences. Nelson (2020) has described the euphoria they experienced from connecting with their peers in the interview context and a resulting sense of contentment with their identity from interacting with others who shared their experience of being queer. Similarly, I believe that exposure to a range of detailed stories from donor-conceived people has normalised my donor-conceived identity. In a similar vein, Mogseth (2019) has argued that:

> Of course, no one had told me, or even insinuated, that I keep my conception a secret, but, then again, no one had told me not to. During my fieldwork, by sharing my experience, and by listening to my interlocutors, the topic of my donor conception lost its inexplicit taboo, and with this loss: a veil of displaced shame dissipated. (p. 115)

Listening to others' stories has also led me to reflect more carefully about the circumstances of my own conception and the ongoing unknowns within that story. Despite the agency and authority that doing research can bring, I am completely powerless over the information I have about my donor, donor 257. I am denied information about my medical and biological history, which I believe is central to the truth about who I am. I continue to wonder if I will be able to find donor 257 before he dies, or whether he will pop up on a DNA testing platform tomorrow. This powerlessness is simultaneously heartbreaking, infuriating and mobilising. The consolation is that I am fortunate to have access to a community of people who know what it feels like to deal with systems which deny us access to information about our own identities and histories. My peers empathise with this struggle and reinforce that I am not alone. On the internet, there is a group of people who are 'like me,' and together we create a space where we belong. I hope that in the future we have more control over the narrative of who we are and what is important to us. Our stories deserve to be told.

Conclusion

'Doing reflexivity' is an ongoing process of considering how my personal attitudes, experiences and motivations affect the decisions I make in my research. This involves exploring how I approach research differently as an insider, as well as contemplating what I need to change or compromise to take on the additional role of researcher. Overall, the opportunity to do research with my peers is a privilege, and I carry the weight of their stories and the responsibility to represent their voices. In this chapter I have reflected on three phases of becoming: becoming donor-conceived, becoming activist and becoming researcher. My trajectory from person with lived experience to researcher reflects a broader intervention in research that recognises that different conclusions are reached by people with lived experience who are embedded in the communities in which they study. People with lived experience bring different priorities and ways of thinking and build lived experience into the ownership of the research process. This coincides with a social shift away from the secrecy and silence of the past, and more than 30 years since the Convention of the Rights of the Child was adopted, there is a strong and solid movement to recognise the rights and best interests of the child and donor-conceived adult that they grow up to be. Major challenges arise in a constantly shifting field where the technologies, industry and legislation are evolving as the people created by the technologies are growing up. This is compounded by the rapid increase in people accessing ART worldwide, the rapid uptake of DNA testing technologies and proliferation of social media platforms. Despite these challenges, legal and policy responses must recognise and build upon donor-conceived people's knowledge, strength and determination. While the role of donor-conceived people is still emerging in research contexts, there is a clear path forward: it is time for researchers to practice meaningful engagement and listen to the voices of donor-conceived people.

Acknowledgements I am indebted to the donor-conceived participants who have generously shared their stories for this research. Sincere thanks to my supervisors Christy Newman, Michele Zappavigna and Kerryn Drysdale for their ongoing guidance and helpful feedback on drafts of this chapter. I would also like to thank the Editor and Reviewers for their comments that improved this chapter. Giselle Newton is supported by an Australian Government Research Training Program Scholarship.

References

Adams, D. (2013). Conceptualising a child-centric paradigm: Do we have freedom of choice in donor conception reproduction? *Journal of Bioethical Inquiry, 10*(3), 369–381. https://doi.org/10.1007/s11673-013-9454-7

Adams, D., & Lorbach, C. (2012). Accessing donor conception information in Australia: A call for retrospective access. *Journal of Law and Medicine, 19*(4), 707–721.

Adams, D. H., Gerace, A., Davies, M. J., & de Lacey, S. (2021). Self-reported physical health status of donor sperm-conceived adults. *Journal of Developmental Origins of Health and Disease, 12*(4), 638–651. https://doi.org/10.1017/S204017442000080X

Allan, S. (2017). *Donor conception and the search for information: From secrecy and anonymity to openness*. Routledge, Taylor and Francis Group.

Allan, S., Adams, D., & Raeymaekers, S. (2020, January 27). Donor-conceived and surrogacy-born children's rights in the age of biotechnology. *BioNews*. https://www.bionews.org.uk/page_147460

Andreassen, R. (2017). New kinships, new family formations and negotiations of intimacy via social media sites. *Journal of Gender Studies, 26*(3), 361–371. https://doi.org/10.1080/09589236.2017.1287683

Attia, M., & Edge, J. (2017). Be(com)ing a reflexive researcher: A developmental approach to research methodology. *Open Review of Educational Research, 4*(1), 33–45. https://doi.org/10.1080/23265507.2017.1300068

Beeson, D. R., Jennings, P. K., & Kramer, W. (2011). Offspring searching for their sperm donors: How family type shapes the process. *Human Reproduction, 26*(9), 2415–2424. https://doi.org/10.1093/humrep/der202

Berger, R. (2015). Now I see it, now I don't: Researcher's position and reflexivity in qualitative research. *Qualitative Research, 15*(2), 219–234. https://doi.org/10.1177/1468794112468475

Blyth, E. (1998). Donor assisted conception and donor offspring rights to genetic origins information. *The International Journal of Children's Rights, 6*(3), 237–253. https://doi.org/10.1163/15718189820494067

Blyth, E. (2002a). Information on genetic origins in donor-assisted conception: Is knowing who you are a human rights issue? *Human Fertility, 5*(4), 185–192. https://doi.org/10.1080/1464727022000199102

Blyth, E. (2002b). Being a child of donor insemination. *BMJ: British Medical Journal, 324*(7349), 1339.

Blyth, E., Crawshaw, M., Frith, L., & Jones, C. (2012). Donor-conceived people's views and experiences of their genetic origins: A critical analysis of the research evidence. *Journal of Law and Medicine, 19*(4), 769–789.

Braun, V., & Clarke, V. (2013). *Successful qualitative research: A practical guide for beginners*. SAGE Publications. https://books.google.com.au/books?id=EV_Q06CUsXsC

Charlton, J. I. (2000). *Nothing about us without us: Disability oppression and empowerment*. University of California Press.

Collins, P. H. (1998). It's all in the family: Intersections of gender, race, and nation. *Hypatia, 13*(3), 62–82. https://doi.org/10.1111/j.1527-2001.1998.tb01370.x

Commonwealth of Australia. (2011). Donor conception practices in Australia. https://www.aph.gov.au/Parliamentary_Business/Committees/Senate/Legal_and_Constitutional_Affairs/Completed_inquiries/2010-13/donorconception/report/index

Crawshaw, M. (2018). Direct-to-consumer DNA testing: The fallout for individuals and their families unexpectedly learning of their donor conception origins. *Human Fertility, 21*(4), 225–228. https://doi.org/10.1080/14647273.2017.1339127

Crawshaw, M., & Marshall, L. (2008). Practice experiences of running UK DonorLink, a voluntary information exchange register for adults related through donor conception. *Human Fertility, 11*(4), 231–237. https://doi.org/10.1080/14647270801908228

Crenshaw, K. (1991). Mapping the margins: Intersectionality, identity politics, and violence against women of color. *Stanford Law Review, 6,* 1241–1300.

Daniels, K. (2020). The perspective of adult donor conceived persons. In K. Beier, C. Brügge, P. Thorn, & C. Wiesemann (Eds.), *Assistierte Reproduktion mit Hilfe Dritter* (pp. 443–459). Springer Publishing. https://doi.org/10.1007/978-3-662-60298-0_29

Dempsey, D., Kelly, F., Horsfall, B., Hammarberg, K., Bourne, K., & Johnson, L. (2019). Applications to statutory donor registers in Victoria, Australia: Information sought and expectations of contact. *Reproductive Biomedicine & Society Online.* https://doi.org/10.1016/j.rbms.2019.08.002. Epub 2019 September 04.

Doyle, C. (2008). *Consumer involvement in dementia care research, policy, and program evaluation.* Australian Institute for Primary Care, 10.

Dwyer, S. C., & Buckle, J. L. (2009). The space between: On being an insider-outsider in qualitative research. *International Journal of Qualitative Methods, 8*(1), 54–63. https://doi.org/10.1177/160940690900800105

Eggins, S., & Slade, D. (1997). *Analysing casual conversation.* Equinox Publishing Ltd.

Few-Demo, A. L. (2014). Intersectionality as the 'New' critical approach in feminist family studies: Evolving racial/ethnic feminisms and critical race theories: Evolving feminisms. *Journal of Family Theory & Review, 6*(2), 169–183. https://doi.org/10.1111/jftr.12039

Fisher, K. R., & Robinson, S. (2010). Will policy makers hear my disability experience? How participatory research contributes to managing interest conflict in policy implementation. *Social Policy and Society, 9*(2), 207–220. https://doi.org/10.1017/S1474746409990339

Flood, M., Martin, B., & Dreher, T. (2013). Combining academia and activism: Common obstacles and useful tools. *Australian Universities Review, 55*(1), 17–26.

Fox, A., & Allan, J. (2014). Doing reflexivity: Moments of unbecoming and becoming. *International Journal of Research & Method in Education, 37*(1), 101–112. https://doi.org/10.1080/1743727X.2013.787407

Frith, L., Blyth, E., Crawshaw, M., & van den Akker, O. (2018). Secrets and disclosure in donor conception. *Sociology of Health & Illness, 40*(1), 188–203. https://doi.org/10.1111/1467-9566.12633

Guillemin, M., & Gillam, L. (2004). Ethics, reflexivity, and 'ethically important moments' in research. *Qualitative Inquiry, 10*(2), 261–280. https://doi.org/10.1177/1077800403262360

Hall, A. E., Bryant, J., Sanson-Fisher, R. W., Fradgley, E. A., Proietto, A. M., & Roos, I. (2018). Consumer input into health care: Time for a new active and comprehensive model of consumer involvement. *Health Expectations: An International Journal of Public Participation in Health Care and Health Policy, 21*(4), 707–713. https://doi.org/10.1111/hex.12665

Halliday, M. A. K. (1978). Language as social semiotic: The social interpretation of language and meaning. London: Edward Arnold.

Hertz, R., & Nelson, M. K. (2018). *Random families: Genetic strangers, sperm donor siblings, and the creation of new kin*. Oxford University Press.

Hewitt, G. (2002). Missing links: Identity issues of donor conceived people. *Journal of Fertility Counselling, 9*, 14–19.

Hodkinson, P. (2005). 'Insider Research' in the study of youth cultures. *Journal of Youth Studies, 8*(2), 131–149. https://doi.org/10.1080/13676260500149238

Jadva, V., Freeman, T., Kramer, W., & Golombok, S. (2010). Experiences of offspring searching for and contacting their donor siblings and donor. *Reproductive BioMedicine Online, 20*(4), 523–532. https://doi.org/10.1016/j.rbmo.2010.01.001

Johnson, A., Lawson, C., & Ames, K. (2018). Are you really one of us? Exploring ethics, risk and insider research in a private Facebook community. *SMSociety '18: Proceedings of the 9th International Conference on Social Media and Society*, 102–109. https://doi.org/10.1145/3217804.3217902

Kirkman, M., & Kirkman, A. (2002). Sister-to-sister gestational 'surrogacy' 13 years on: A narrative of parenthood. *Journal of Reproductive and Infant Psychology, 20*(3), 135–147. https://doi.org/10.1080/02646830276 0270791

Knight, N. K. (2010). *Laughing our bonds off: Conversational humour in relation to affiliation*. (PhD Thesis), University of Sydney. http://hdl.handle.net/2123/6656

Kohl, E., & McCutcheon, P. (2014) (2015). Kitchen table reflexivity: negotiating positionality through everyday talk. *Gender Place & Culture, 22*(6), 747–763. https://doi.org/10.1080/0966369X.2014.958063

Macmillan, C. (2016). *A study on the effects of donor conception, secrecy and anonymity, according to donor-conceived adults*. (MA Thesis), Macquarie University. http://hdl.handle.net/1959.14/1262038

Mahlstedt, P. P., LaBounty, K., & Kennedy, W. T. (2010). The views of adult offspring of sperm donation: Essential feedback for the development of ethical guidelines within the practice of assisted reproductive technology in the United States. *Fertility and Sterility, 93*(7), 2236–2246. https://doi.org/10.1016/j.fertnstert.2008.12.119

Martin, J. (2009). Genre and language learning: A social semiotic perspective. *Linguistics and Education, 20*(1), 10–21. https://doi.org/10.1016/j.linged.2009.01.003

Mason, J. (2006). Mixing methods in a qualitatively driven way. *Qualitative Research, 6*(1), 9–25. https://doi.org/10.1177/1468794106058866

Mauthner, N. S., & Doucet, A. (2003). Reflexive Accounts and Accounts of Reflexivity in Qualitative Data Analysis. *Sociology, 37*(3), 413–431. https://doi.org/10.1177/00380385030373002

Mogseth, M. E. (2019). *Donor conception and unknown kin: Reconsidering identity and family through anonymous and deanonymized relations*. (MA thesis), University of Oslo. https://www.duo.uio.no/handle/10852/69613

Nathan, S., Newman, C., & Lancaster, K. (2018). Qualitative interviewing. In P. Liamputtong (Ed.), *Handbook of research methods in health social sciences* (pp. 1–20). https://doi.org/10.1007/978-981-10-2779-6_77-1

Nelson, R. (2020). Questioning identities/shifting identities: The impact of researching sex and gender on a researcher's LGBT+ identity. *Qualitative Research, 20*(6), 910–926. https://doi.org/10.1177/1468794120914522

Nelson, M. K., Hertz, R., & Kramer, W. (2013). Making sense of donors and donor siblings: A comparison of the perceptions of donor-conceived offspring in lesbian-parent and heterosexual-parent families. In P. N. Claster & S. L. Blair (Eds.), *Contemporary perspectives in family research* (Vol. 7, pp. 1–42). Emerald Group Publishing Limited. https://doi.org/10.1108/S1530-3535(2013)0000007004

Newman, J., Paul, R., & Chambers, G. (2020). *Assisted reproductive technology in Australia and New Zealand 2018*. National Perinatal Epidemiology and Statistics Unit, the University of New South Wales, Sydney. https://npesu.unsw.edu.au/sites/default/files/npesu/data_collection/Assisted%20Reproductive%20Technology%20in%20Australia%20and%20New%20Zealand%202018_0.pdf

Porter, J., Parsons, S., & Robertson, C. (2006). Time for review: Supporting the work of an advisory group. *Journal of Research in Special Educational Needs, 6*(1), 11–16. https://doi.org/10.1111/j.1471-3802.2006.00055.x

Rodino, I. S., Burton, P. J., & Sanders, K. A. (2011). Donor information considered important to donors, recipients and offspring: An Australian perspective. *Reproductive BioMedicine Online, 22*(3), 303–311. https://doi.org/10.1016/j.rbmo.2010.11.007

Rose, J. (2009). *A critical analysis of sperm donation practices: The personal and social effects of disrupting the unity of biological and social relatedness for the offspring* (PhD Thesis), Queensland University of Technology. https://eprints.qut.edu.au/32012/

Schneider, B. (2012). Participatory action research, mental health service user research, and the hearing (our) voices projects. *International Journal of Qualitative Methods, 11*(2), 152–165. https://doi.org/10.1177/160940691201100203

Shaw, R. M., Howe, J., Beazer, J., & Carr, T. (2020). Ethics and positionality in qualitative research with vulnerable and marginal groups. *Qualitative Research, 20*(3), 277–293. https://doi.org/10.1177/1468794119841839

Smith, C. B. R. (2016). 'About nothing without us': A comparative analysis of autonomous organizing among people who use drugs and psychiatrized groups in Canada. *Social Work, 5*(3), 28.

Song, M., & Parker, D. (1995). Commonality, difference and the dynamics of discourse in in-depth interviewing. *Sociology, 29*(2), 241–256.

Taylor, J. (2011). The intimate insider: Negotiating the ethics of friendship when doing insider research. *Qualitative Research, 11*(1), 3–22. https://doi.org/10.1177/1468794110384447

Ti, L., Tzemis, D., & Buxton, J. A. (2012). Engaging people who use drugs in policy and program development: A review of the literature. *Substance Abuse Treatment, Prevention, and Policy, 7*(1), 47. https://doi.org/10.1186/1747-597X-7-47

UNICEF. (1989). *Convention on the rights of the child*. Human Rights Office of the High Commissioner. https://www.ohchr.org/en/professionalinterest/pages/crc.aspx

Whipp, C. (1998). The legacy of deceit: A donor offspring's perspective on secrecy in assisted conception. In E. Blythe, M. Crawshaw, & J. Speirs (Eds.), *Truth and the child 10 years on: Information exchange in donor assisted conception*. British Association of Social Workers Publications.

World Health Organization. ([1978] 2004). *Declaration of Alma Ata: Report of the international conference on primary health care, Alma-Ata*, USSR, 6–12 September 1978. *Development, 47*, 159–161. http://link.springer.com/10.1057/palgrave.development.1100047

13

Reproductive Choices and Experiences in Planning for Parenthood and Managing Infertility

Sonja Goedeke, Maria Mackintosh, and Lara Grace

Within many societies, motherhood has traditionally been constructed as a normal, biological, and inevitable part of a woman's identity, particularly within the context of the heteronormative, nuclear family (DeSouza, 2013; Donath, 2015; Perrier, 2013). While this pronatalist discourse has been challenged and disrupted by women's access to education and employment, the availability of contraception and legalisation of abortion in many countries, all of which have given women increased reproductive autonomy (Cooke et al., 2012; Lavender et al., 2015), there remains a social expectation that women become mothers. Indeed, in Aotearoa New Zealand, three out of four women become mothers at some point in their lives, even if, as is the case both internationally and domestically, this is delayed, and the age of women having their first children has shifted to thirty and above (Cooke et al., 2012; Statistics New Zealand, 2018). Research in Australia and Aotearoa New Zealand, reflects

---

S. Goedeke (✉) • M. Mackintosh • L. Grace
Auckland University of Technology, Auckland, New Zealand
e-mail: sonja.goedeke@aut.ac.nz

© The Author(s), under exclusive license to Springer Nature Singapore Pte Ltd. 2022
R. M. Shaw (ed.), *Reproductive Citizenship*, Health, Technology and Society, https://doi.org/10.1007/978-981-16-9451-6_13

on social, media, and political discourses valuing motherhood, and how these may lead to experiences of exclusion, stigma, and invalidation in the personal and professional lives of women without children (Tonkin, 2019; Turnbull et al., 2017). Such experiences may contribute to women's sense of childlessness as deeply distressing, influence their subjective wellbeing and quality of life, and affect their sense of identity as women (Shreffler et al., 2020).

Yet at the same time as motherhood continues to be valued as a fundamental role for women, DeSouza (2013) argues that it is a particular type of motherhood—an individualised form of mothering—in contrast with the more shared child-rearing practices of other societies, which has been promoted by the dominant western middle class in societies such as Aotearoa New Zealand. This can be argued to have consequences both for which groups of women are encouraged to be mothers, and their subsequent experiences. DeSouza highlights how values of individualism charge women with ever-increasing responsibility for maximising the moral, social, and psychological development of their children. Similarly, international authors have reflected on a shift across generations in Western societies to what has been termed a "good mother" or "intensive mothering" discourse, which sets high expectations for mothers and encompasses a belief that motherhood is child-centred, time-intensive, emotionally engrossing, and characterised by fulfilment (Giesselmann et al., 2018; Matley, 2020; Nomaguchi & Milkie, 2020). Hays writes that intensive motherhood, "requires the day-to-day labour of nurturing the child, listening to the child, attempting to decipher the child's needs and desires, struggling to meet the child's wishes, and placing the child's well-being ahead of their [mothers'] own convenience" (Hays, 1996, p. 115). A belief in parental determinism, that how children turn out will depend largely on parenting (Nomaguchi & Milkie, 2020), appears to underlie such approaches and lead to the investment of 'great swaths of time, money, energy and emotional labour' into raising children (Elliott et al., 2015, p. 352). While feminist models challenge the centrality of motherhood for women, many researchers have argued that expectations of motherhood have only intensified since Hays (1996) wrote about them, and that in Western societies there is pressure for women to be "perfect mothers" (Henderson et al., 2016; Nomaguchi & Milkie, 2020).

The implications of such individualised and intensive, child-centred mothering ideology have been studied across a range of contexts, including on women's mental health and wellbeing (Giesselmann et al., 2018), their experiences of combining careers and parenting (Christopher, 2012), parenting decisions such as breastfeeding (Hanser & Li, 2017), and their ability to express maternal regret (Matley, 2020). While some have argued that these beliefs around good mothering rely on access to wealth and are less relevant for those living in poverty or with disabilities (Mackendrick, 2014), these discourses also need to be negotiated by mothers across different demographics (Elliott et al., 2015), particularly in the context of dominant cultures (Le Grice et al., 2017). Indeed, DeSouza (2013) draws attention to how such discourses may serve to encourage motherhood for certain groups of women (largely the white, middle class), while women whose subjectivities have been formed outside of white, middle-class, or Western contexts (such as immigrant and indigenous women) may be discouraged from doing so, and/or may subject them to discrimination and even unresponsive care from health professionals.

Given these social and cultural trends around motherhood, in this chapter we explore how women make decisions about becoming parents and what experiences they may have when their parenting plans are not achieved, such as when they experience infertility. We draw on research from three qualitative research studies. In the first, young women's views and understandings regarding the 'right' time for motherhood in Aotearoa New Zealand were explored through conducting two focus groups with 13 women aged 25 to 32 years old. Note that the majority identified as heterosexual and New Zealand European/Pākehā, were in relationships, held paid employment, and at a minimum had undergraduate qualifications. In the second and third studies, six women's experiences of childlessness and six women's experiences of adjustment to parenthood following fertility treatment were explored through semi-structured interviews. All women were aged between 30 and 50 years and again, identified mainly as New Zealand European/Pākehā. Transcribed data from the focus groups and interviews were analysed drawing on Braun and Clarke's well-established six-step process for analysing qualitative data. We begin the chapter by drawing on the first study and describing the conditions that these women deemed necessary to fulfil to become 'good' parents, and then, drawing on the other two studies, describe how women's experiences of infertility and parenthood and their decisions around treatment are shaped by individualistic and 'good mother' ideologies, and how they work to transform and resist these discourses.

While our work reflects the experiences of mainly Pākehā, middle class and heteronormative women, we use their experiences to discuss how women's reproductive decision-making and their experiences related to becoming parents need to be understood from a holistic perspective that recognises the impact of biological, social, cultural, and structural factors on their experiences and how these may constrain their reproductive autonomy. We thereby also draw attention to the potential impact of intensive mothering and how its valorisation occurs at the expense of other forms of mothering and being, including in Aotearoa New Zealand.

Planning for Motherhood

Previous research has suggested that women make decisions about their reproductive timing informed by normative narratives defining the "right" time for motherhood (Donath, 2015; Gotlib, 2016). Dominant within Western societies are narratives suggesting that the appropriate time is when women are not "too young" or "too old," and when certain personal and social milestones have been achieved (Lavender et al., 2015; Perrier, 2013). While intensive mothering has been ascribed to white, middle-class women (Lareau, 2011), as noted above, different groups of mothers within Western societies, such as those from socio-economically disadvantaged backgrounds, marginalised ethnicities, and those who are single mothers, are also affected by the ideology of intensive mothering with implications for societal approval of their position as parents and for their parental wellbeing (Dotti Sani & Treas, 2016; Elliott et al., 2015; Ishizuka, 2019). For example, Graham (2018) outlines how young Māori mothers have been pathologised and denigrated widely through various studies and the media, and characterised as the least suitable and desired mothers, constituting social problems (e.g., see Chap. 7 in this volume with reference to this point). Participants in the focus groups in our study were predominantly Pākehā, economically advantaged, and university educated. They made their decisions around the 'right' time for motherhood with reference to the 'right' time not just to become *a* mother, but a '*good*' mother in line with the tenets of the 'intensive mothering' ideology to which they both ascribed, and which was seemingly reinforced by social pressures from family, friends, and the media. We acknowledge that their decisions are influenced by their position of relative privilege and reflect that their experiences of intensive mothering ideology influence not only the subsequent conditions they felt pressured to achieve to position themselves for motherhood, but are also classed and raced, and thereby undermine and disadvantage other iterations of mothering. Graham (2018), for instance, has highlighted how setting Pākehā experiences as the benchmark for Māori wellbeing may not be standpoints that encompass the lived and contextual realities of all mothers.

The 'Right' Age to Make a 'Good' Mother

Similar to previous studies (Bell, 2013; Perrier, 2013), participants in our study associated younger mothers with a lack of emotional and financial stability and older mothers with increased health problems later in their children's lives. Consequently, participants connected both younger and older mothers with reduced abilities to be 'good' mothers and as outside the ideal time for motherhood. In particular, our participants, presumably because they were at an age where being a younger mother was no longer applicable, questioned whether having children as older mothers was a responsible parenting decision. This included reference to older mothers depriving their children of the opportunity to have grandparents and of grandparents being unavailable to support parents in being 'good' mothers. Our participants' views align with a "good" mother discourse, which by extension suggests that contemporary grandparenthood is similarly about "being there" for your children and your grandchildren (Perrier, 2013). Participants, however, also emphasised the significance of achieving personal and relational milestones before motherhood to enable them to become good parents.

The 'Right' Personal and Relationship Factors to Be a 'Good' Mother

Participants discussed the achievement of middle-class lifestyle milestones, such as finishing education, establishing careers, developing financial security and being in a committed relationship, as related to the 'right' time for motherhood. As reported in other research (Cooke et al., 2012; Perrier, 2013), these may be related to women's reasons for delaying motherhood.

Perrier (2013) suggests that delaying motherhood until these conditions are satisfied helps women secure a moral position, one in which they have reached the 'right psycho-social time' to become 'good' mothers. In our study participants drew connections between financial stability and being 'good' parents as is reflected in an interaction where Lisa talks about the questions she asks herself to assess her readiness for parenting, including 'do I have the money to bring someone else into this world and be able to support them?', and Belle adds, 'to give your child the *best* life?' Financial stability was also valued by our participants since it was constructed as making it possible to be stay-at-home parents, which was viewed as having significant benefits for child development. Tina reflects that it would be ideal if, 'she would get to spend more time at home in those early years, which … are quite important … as they grow up.' Similarly, Anna comments that, 'it comes down to money, and that sucks, because that impacts on a child's wellbeing … to be there in the formative years.'

Related to financial stability was the perceived importance of having established a career prior to motherhood, which was seen as increasing the likelihood of providing women the option to take time out from their careers to care for their children without significantly impacting their career progression. Career progression and parenting responsibilities were often positioned as in conflict, however, as reflected by Lisa, who commented, 'I think if I had kids, I'd have to sacrifice some of that. I probably wouldn't do my job as well because I'd have other priorities.'

Previous studies (Bass, 2015; Ussher, 2015) reported similar findings with both roles (career and mother) perceived as time-intensive, and with motherhood seen as jeopardising career development. Most women in the current study valued their careers but drew a connection between developing careers prior to motherhood and being 'good' mothers. Such findings reflect a tension between neo-traditionalist economic-nurturing ideologies and feminist models, with their emphasis on women finding fulfilment outside the home and caregiving responsibilities.

In her discussion of Intensive Mothering, Hays (1996) argued that mothers follow contemporary models unconsciously, perceiving expectations of motherhood as natural and necessary. Henderson et al. (2016) similarly suggest that even when women do not consciously subscribe to intensive mothering ideologies, their influence is inevitable. Throughout our participants' accounts, motherhood was constructed as a huge responsibility, requiring women to possess 'emotional maturity', defined as self-awareness and emotional stability alongside a willingness to make sacrifices in meeting the responsibilities of motherhood. For example, Natalie commented that she needed to be, 'ready to be a mother psychologically. I think it really affects how you treat your children, really affects who they become so it's really important.' In contrast, mothers perceived as lacking emotional maturity (e.g., having children to satisfy their own needs) were constructed as selfish, irresponsible, and likely to cause negative implications for their child. Anna and Rebecca reflect critically on the notion that children may be 'born with jobs … like to heal a relationship … or yourself' and ponder 'how this is going to affect the child?'

Interestingly, many of our participants, the majority of whom identified as heterosexual, positioned themselves as holding the responsibility for planning children, thereby conforming to more traditional roles of femininity and masculinity, and the implicit positioning of women as responsible for children in the intensive mothering discourse. Although our participants did not believe that childcare should exclusively be undertaken by mothers, 'good' mothers were constructed as those that plan for parenthood carefully, put aside their own needs and desires, and are fully committed to their roles. Anxiety around their ability to meet the standards required to become "good" mothers created pressure and, similar to the women in Maher and Saugeres' (2007) study, women discussed postponing motherhood until they felt they were in a position to be "good" mothers.

While the planning to become parents seemed vested mainly in the women, a joint parental desire and a committed relationship were constructed as necessary for motherhood. Numerous other studies suggest that a stable relationship and a shared choice are regarded as key conditions for parenthood (Bell, 2013; Cooke et al., 2012; Martin, 2017). What is interesting is that in the current study, relationship status and the shared desire for children was discussed in relation to quality of parenting, with participants reflecting on potentially negative outcomes for children if women decided to have children in the absence of their partners' agreement. Rebecca, for example, reflected that there is a 'flow on effect of that (for children), with all of these issues around identity, being loved, am I good enough?' and suggests that where the joint desire is absent, 'It's actually quite irresponsible of people to do it, because it's going to have such a major effect on how you're parenting.'

Social Pressures Related to Being a "Good" Mother

Many participants reported that their families regularly pressured them to have children because they were perceived to be at a certain age or stage in their life where motherhood was expected. Participants also pointed to the role of the media, suggesting that unrealistic portrayals of a 'good' mother often served to reinforce parenting standards perceived to be unreachable, potentially delaying their decisions to become mothers. For example, Natalie reflected how media portrayals showing, 'a happy woman, perfectly dressed, with a very happy child' related to her reluctance to become a mother before she felt she could achieve this ideal.

This line of thinking is supported by other studies (Cooke et al., 2012; Lavender et al., 2015; Martin, 2017), although Henderson et al. (2010) identified that it is not only the media that helps to sustain this unrealistic view of a 'good' mother, but also mothers themselves, specifically in their interactions with other women. This was congruent with the findings from our study where participants discussed how this ideal of a 'good' mother was reinforced by friends, partly through social media, in which women typically shared examples of their mothering experiences reflective of intensive parenting. One participant reflected on how 'possibly most of what I see is all these really positive photos, and it's this wonderful time, and how happy they are' (Lisa).

Finally, participants believed that employers would be biased against mothers. As Tina said, 'There's that stigma, I think, when you do go for jobs, that if they know that you have children, that if your kid is sick, you'll have to stay home.' These perceptions are supported by research findings indicating that when equally qualified female candidates were evaluated for jobs those who were mothers were perceived as less committed to the job than non-mothers (Correll et al., 2014).

Implications of Pursuing the "Right" Conditions to Be a "Good" Parent

Similar to existing studies (Daniluk, 2015; Martin, 2017; Perrier, 2013), our research highlighted the tension women may experience between wanting to satisfy personal and relational conditions before motherhood, while also wanting to have children before they are "too old." For example, most of our participants planned to delay motherhood until their early to mid-thirties in anticipation of achieving specific life milestones, but this age coincides with a decline in female fertility and may result in difficulties in having children. These findings illustrate an incompatibility between reconciling the perceived 'right psychosocial time' with the 'right biological time' to be considered a 'good' parent.

This was further complicated by our participants' limited knowledge regarding the age at which female fertility *significantly* declines, an overly optimistic view of the accessibility and effectiveness of assisted reproductive technologies and limited knowledge of the increased adverse pregnancy outcomes associated with advanced maternal age. The women in our study were thus making decisions to delay motherhood to align mainly with the 'right psychosocial time' to become 'good' mothers, without completely understanding the possible consequences of their decisions. Other studies (inclusive of one in Aotearoa New Zealand) have similarly identified that the majority of people lack a thorough understanding of fertility and are overly confident about women's chances of having children at delayed maternal ages (Lucas et al., 2015; Peterson et al., 2012).

These findings highlight the need for interventions to go beyond the individual level and target the social and cultural contexts which may constrain women's reproductive decisions. This includes accessible education about fertility, support for women in managing career progression and family life responsibilities through improved family-friendly policies (Bass, 2015; Ussher, 2015), and more equitable distribution of caregiving and domestic responsibilities (as current research suggests that women continue to be primarily responsible for childcare and domestic activities (Coombe et al. 2019)). Similarly at a government level, there is a need to ensure appropriate parental leave policies, family allowances, and accessibility of childcare options, provisions which our participants deemed insufficient in Aotearoa New Zealand. Finally, however, it is also important to critique Western, middle-class norms surrounding individualised constructs of motherhood and the "right" time for motherhood, challenging parenthood as an individual endeavour and challenging the need to have "everything just right" in order to fulfil the expectations of intensive parenting (Daniluk, 2015; Lavender et al., 2015; Martin, 2017). As DeSouza (2013) has argued, such ideologies advantage those groups of people who have access to the resources perceived necessary to embark on parenting as "managed projects requiring assessment/research, planning and implementation skills" (p. 18) and may subsequently deny mothers who do not meet these conditions and position them as unable to be "good" mothers. Similarly, Graham (2018) challenges the deficit discourses that stem from these ideologies, and through her study of young Māori mothers demonstrates how what wellbeing ('good' motherhood) means must be understood within the context of a Māori worldview. Furthermore, our study illustrates how these expectations can have negative consequences both for when motherhood is elusive and for parenting, as we discuss next.

When Motherhood Is Elusive

The experience of infertility is frequently described in the literature as emotionally challenging, as stressful, as involving significant loss and grief, and as having negative effects on identity (Shreffler et al., 2020). These experiences were similarly reflected in our study of six women, all of whom had attempted to become mothers using assisted reproductive technologies, including between three and six cycles of IVF, but without success. Their experiences appeared compounded by a backdrop of pro-natalist and 'intensive motherhood' ideologies, since these position motherhood as an all-encompassing role (with little room for other options for women), and one for which women need to plan and be well-prepared in order to be 'good' mothers. The transition to a childfree existence was experienced as difficult in this context, and complicated by either a self-held or other-held assumption that the women may not have met the conditions required for 'good mothering,' further complicating their distress.

Motherhood as Central to Identity as Women

All of the women in our study had anticipated that becoming a mother would be an irrefutable part of their life plan. This loss of anticipated 'motherhood' identity was experienced as an identity shock and was reflected in the way in which many described feelings of failure as women. Christina commented, 'I've kind of failed … as a woman, these are fundamental things … getting married and having a baby.' Isabelle reflected how motherhood was something she had always wanted, saying, 'I think the hard thing for women is that when you are a little girl you play with dolls and you never envisage that you're *not* going to be a mum.' She reflected on her distress in the context of, 'everything in this society screams: families, children, it's the way to be a woman', and how 'you're surrounded by the babies and the prams and the advertising. You are not even safe in your own home. It's everywhere' and how she had the expectations of 'not only our family, but our parents and our siblings and the extended family' on her to have a child.

For most of our participants there was a loss of meaning associated with this loss of identity. April commented, 'Our family is shallow … there's no depth to it.' Motherhood was seen as giving life purpose, and without it, women struggled to renegotiate a different purpose. Callie spoke of how 'my life had been put in limbo. I was in this no-man's land of not knowing what to do with my life,' and working out, 'what is our purpose going forward if it's not to raise children?'

Letherby (2002) highlights that individuals who are infertile experience a loss of a desired identity which is distressing in itself but may also lead to isolation and disconnection from others. In our study, Callie discussed how there were three 'tribes' that being infertile meant she did not belong to; those still undergoing treatment, those with children, and those who never wanted to have children—reflecting a very real disconnect from other women. Similarly, Isabelle spoke of how, 'I'm sort of on the outside of a club or society that everybody else knows what that's all like and I have no idea.' She expanded that, 'the implications of not having kids at all is actually wider than what people think.' Women reflected that this disconnect was envisaged to be long-term, for example, talking of how they would not get to experience being mothers-in-law or having grandchildren.

Other research has similarly highlighted the social isolation that women may experience during treatment, how this often persists post-treatment, and that there is a lack of information available on how child-free women can live a fulfilled life (Wirtberg et al., 2007). We would argue that this experience may be intensified in the context of values of 'intensive mothering' which depict biological motherhood as central and all-consuming. Callie, for example, spoke of the 'suffering' and 'massive grieving' that came with infertility, and the majority of women described pain that was ongoing, as 'cumulative … cycles of grief that continue' (Callie), that are 'actually going to be part of who I am and my journey in life forever' (Isabelle).

Investment in Being "Good" Mothers

Women in our study spoke of how they had actively planned for their reproductive choices, or as Isabelle described it, how she had 'focused my whole life in on trying to have this baby.' For women this included ensuring healthy lifestyles and considering all available treatment options entailing significant investment and effort on their part. April commented that, 'if somebody had said to me "look, if you spun on your head and you drunk milk through your nose and you wore purple poke-a-dot knickers, while you did it," then I'd do it.' Christina reflected how she felt an imperative to do everything she could, because if she did not, that would imply that she was not being the greatest mother.

Their experiences were made additionally distressing through the perceived ease with which some around them had been able to procreate, and their sense of resentment and injustice when these individuals were ones who had *not* planned for motherhood and were considered to be 'undeserving' or less than ideal candidates to make 'good' mothers. Some participants also pointed to assumptions made by others in regard to their childlessness, including that they had, 'obviously put (their) careers first' (Isabelle) that, 'women were leaving it too late and concentrating on their careers' (Arizona) and that, 'you shouldn't have left it that long' (Callie). There was anger and guilt in response as they felt they had done everything 'by the book' to become good mothers, but in so doing, were positioned as either "mad" in their attempts ('I turned into a crazy person', Callie) or "bad" and to blame for their childlessness. Indeed, Ulrich and Weatherall (2000) have written how women with infertility may sometimes be portrayed as desperate and willing to go to extraordinary, sometimes irrational lengths in their quest to have children, or especially if they are older that they have engaged in lifestyle choices that have affected their fertility; the "fault" is thus assigned to them (Payne & Goedeke, 2009). In this way, those with fertility issues may be positioned as "mad, bad or sad" (McLeod & Ponesse, 2008; Payne & Goedeke, 2009).

Interestingly women's decisions to discontinue attempts to become a mother were often made in consideration of factors deemed relevant to being a 'good' mother. This included medical advice to discontinue treatment because of a perceived risk to themselves or future children, or advancing age which set material constraints on their ability to conceive and was considered in light of the implications for offspring. One participant, Isabelle, discussed having an 'ethical responsibility' to do 'what's best for a child' and reflected that while the technology might allow her to continue to try and have children, 'I think the thing that people don't really quite get is that sometimes your decision-making is not really around what you personally want or what you would like to happen but probably more what's best for a child.' For her, pursuing further treatment to have a child would place her in a position of not being 'responsible.' Similarly, Arizona reflected on her decision-making process, which included pondering, 'am I doing this for the right reasons? Is this the right thing for a child?'

Moving On

Part of women's ability to forge identities as women without children was based on their decisions to discontinue attempts in the context of being assured that they had done everything possible to become mothers, and that their decisions were based on what would be best, not only for them, but also future children. For a few of the women, there came a point in their journey in which they experienced relief at discontinuing treatment and of no longer feeling controlled by their desire to achieve the unachievable. Matthews and Matthews (1986) define this as a reality reconstruction whereby the reality of infertility and being childfree becomes incorporated into the individual's identity. To do this, individuals needed to acknowledge the losses associated with remaining childfree while rejecting the socially constructed link between fertility and self-worth. For some this involved consciously considering what children would have meant to them, asking 'what are the reasons we actually want children?' (Meredith) and 'is it just because it's this urge and everyone else is doing it?' Others were also able to consider the negative aspects of having children. Christina reflected on this, saying, 'You know, having children, raising children in this world, in this time we live in, I think it's so difficult.'

In addition, these women were able to embrace characteristics of themselves and possibilities they previously had not considered and were able to find positive aspects to their identity. Isabelle reflected that it was about asking, 'You're not a mum, but you know, what do you want? What do you want your life to be?' For some this was about new projects and other ways to find 'meaning in their life and things that [they] can do that's productive, that's creative, that contributes to society' (Isabelle). These comments highlight the centrality participants had accorded to motherhood in their lives, and the nuclear context within this was positioned. One participant commented that in retrospect making a child the focus of her life may not have been right for her, or, interestingly, for a child. She remarked, 'this would (have) be(en) my life for at least, 14 years you know. Um, focusing around this child, and I don't know if that would've been right for me. Or more particularly the child' (Arizona).

In Wirtberg et al.'s (2007) study of adjustment following infertility, the majority of women discussed accepting their "non-parent" identity. Similar to the women in the current study, they had been able to embrace the advantages of living childfree by investing in themselves and focusing on other alternatives. As one participant in our study expressed, 'you have to find some good in the bad. And that's what we've done' (April). Another stated that, 'it's a gift in learning that not always everything always goes the way you want it to and sometimes you have to change your plans or look at other ways of dealing with life' (Isabelle). Most participants expressed a similar sentiment but emphasised that their nonparent identity and focusing on other aspects had been difficult, and a conscious choice.

This process was facilitated for the women in our study, as in previous research (Malik & Coulson, 2008), through being supported by others who had gone through similar experiences. This helped in lessening feelings of isolation, normalising, and reducing the stigma of living 'childfree', and establishing a new childfree identity. This highlights the importance of having other models of womanhood available to women in their reproductive decision-making, allowing women to challenge and resist dominant ideologies. Nonetheless, our study highlights how this transition was a challenging one, complicated by pronatalist discourses and intensive mothering ideologies which frame motherhood as the central role for women and positing certain conditions as needing to be fulfilled in order to be able to assume roles as mothers.

Becoming a Mother After Infertility

Some authors propose that because of the psychological impact of infertility and its treatment, women may have unresolved feelings which may adversely affect their experiences as they transition to pregnancy and parenthood (Fisher et al., 2005; Maehara et al., 2020). For example, research has suggested that women who conceive through fertility treatment experience more pregnancy-specific anxiety and ambivalence and are more at risk for depression and other difficulties in adjusting to parenthood than women who conceive naturally (Maehara et al., 2020). Similarly, Fisher et al. (2005) reported that admissions into an early parenting programme were higher for women who had conceived through reproductive technologies, suggesting that conception via fertility treatment may be associated with an increased rate of early parenting difficulties. In addition, research has suggested that women who conceive via fertility treatment report lower self-esteem in early parenthood and less parenting competence in comparison to mothers who have conceived naturally (Allan et al., 2019; Gibson et al., 2000; McMahon, Ungerer, Tennant, & Saunders, 1997a).

In contrast, women in a longitudinal study by Hjelmstedt et al. (2004) reported that they had been able to put their fertility journey behind them as they transitioned into parenthood, and that it had only a limited emotional impact on their lives. Further, McMahon, Ungerer, Beaurepaire, et al. (1997b) found no differences between mothers who had conceived naturally and those who had conceived via treatment in measures of anxiety, marital satisfaction, postnatal depression, and use of support services. Fisher et al. (2008) suggest that achieving a successful result after considerable psychological, physical, and social stress may actually result in women viewing the transition to motherhood as unproblematic and rewarding.

While such results are optimistic about women's adjustment following treatment, it is possible that results stem in part, from women feeling less able to express doubts or mixed feelings about the realities of motherhood (Fisher et al., 2008). Others (e.g. Gibson et al., 2000; McMahon, Ungerer, Tennant, & Saunders, 1997a) further suggest that IVF mothers place very high parenting expectations on themselves and that it is these that may account for negative early postpartum adjustment differences between IVF and naturally conceiving parents. Indeed, research by Lee et al. (2012) demonstrated that societal-orientated parenting perfectionism was associated with poorer parental adjustment, higher parental stress, and lower parenting self-efficacy and satisfaction among new mothers and fathers. Several other studies have confirmed similar results, identifying an increase in negative mental health consequences for parents who endorse intensive parenting ideologies (Gunderson & Barrett, 2017). In relation to motherhood specifically after infertility, Mohammadi et al. (2014) identified "super-mothering" following experiences of fertility treatment in mothers in Iran and described mothers who wanted to be exceptional in their mothering responsibilities without fatigue.

In our third study, we thus explored how six women adjusted to parenting following successful treatment. All had used assisted reproductive technologies to conceive and had had their children within approximately 18 months of treatment. Children were aged between 6 and 18 months at the time of the study, and women were aged between 29 and 39.

Pregnancy, Birth and Early Parenting: Preparing for the Best, but Expecting the Worst

Women in our study commented on how they had done everything they could to ensure a positive outcome from their pregnancies. Alexis commented on how she had gone, 'to the absolute extreme of being of cautious,' doing her, 'absolute best to make sure that this baby is healthy.' Similarly, the women reflected on having done everything to ensure a safe birth. Nonetheless, women reported being anxious about the pregnancy and birth, with Stevie commenting that during her pregnancy she 'wasn't letting (her)self get attached (to the unborn child) at all … because (she) was terrified that something would happen.' Participants in our study almost seemed to expect something to go wrong, and this operated as a defence mechanism, with Moira commenting 'it's better that you expect the worst. And if the worst doesn't happen it's a win.'

At times this extended past the birth and into early parenting. Stevie, for example, explained that, 'The first six weeks I actually just spent feeling terrified again that something would happen to him. Or that he would be taken off me, that I'd just wake up one morning and he wouldn't be there. Or he wouldn't be breathing.' Some of these fears appeared to affect early parenting, for example, for Wendy who said that the 'bond (between her and her child) wasn't as instant' as she thought it might be for others who had not utilised assisted reproduction. Most of the women in our study however reported that the anxiety they had around pregnancy and birth did not spill over into their parenting longer term or affect their overall adjustment.

Adjustment to Parenting

Indeed, as in some other research (e.g., Hjelmstedt et al., 2004), the women in this study had positive views on their adjustment, which could partly be attributed to their successful result, and possibly, as is suggested above, constitute a minimising of distress associated with parenting, especially given their parenting was 'hard won' and that this might make it difficult to express any uncertainties. However, it is also possible that their positive adjustment could be related to their willingness to be open about the difficulties associated with their reproductive journeys, including those related to parenting, and to the role of social support.

The women in our study were generally open about their struggles, at least with a selected group or over time, and spoke of how sharing experiences with other mothers throughout their journeys, for example in fertility and parenting support groups, was found to be helpful. Moira reflected that she was 'very, very open' and Alexis emphasised that her journey 'shouldn't just be hidden away.' Research has suggested that social support has a protective effect over the emotional distress of fertility treatment (Malik & Coulson, 2008). Support groups are held to be particularly valuable as not only do they facilitate the exchange of common experiences with individuals who have a first-hand understanding of those same experiences (Malik & Coulson, 2008) but they also reduce the stigma women may experience and perceptions of being 'other' or 'defective.' Twyla captures this when she said, 'having a couple of people who had gone through the same thing made you feel a bit more normal.' It is possible that having had the experience of being open about their fertility struggles and having had a child in a way that differed from the expected, facilitated the ability to be open about the difficulties of parenting also, difficulties to which our participants readily admitted. Indeed, Twyla commented that she had given another woman undertaking treatment the advice that, 'just because you've had fertility treatment doesn't make it any easier and you don't have to enjoy it every single second of the day like you thought you would when you didn't have a child.' Participants were reassured that their 'experience (was) probably the same as (for) anyone who is becoming a mother for the first time' (Moira). The women in our study thus drew strong parallels between themselves and other women and were able to do so in the context of connection with other parents or parenting support groups, either online or in person. Wendy reflected that by talking with other mothers, she found that, 'everybody had a really difficult time,' which was 'good to know, good to hear.' Parenting struggles thus became normalised and validated through discussions with other mothers, which helped increase their feelings of belonging, decreased their 'other' identity, and facilitated adjustment. Women also found that even in the 'standard' (rather than fertility-based parent support groups) there were a number of women who had had fertility treatment, and that this was 'quite common' (Moira/Alexis).

In recent times, online support groups and "mommy blogs" have emerged as a form of activism through which women can express themselves and share the joys as well as constraints of motherhood with an audience, thereby resisting and reframing aspects of "intensive mothering" discourses (Matley, 2020; Pedersen, 2016; Petersen, 2015). Allan and Finnerty (2007) suggest that it is those parents who have integrated their fertility history into their lives who have improved emotional well-being, whereas those who have not, experience higher levels of shame, anxiety, and sadness. In our study, we suggest, women who had children following fertility treatment adjusted well not only because they had children, but also because they were able to resist and reframe dominant discourses of perfect mothering in the context of support in their social networks, including fertility support and parenting groups. These facilitated an acceptance of both less than perfect or traditional conditions and pathways towards parenting (IVF journeys) and of parenting itself. As Henderson et al. (2016) point out, reframing intensive parenting culture is not as simple as mothers lowering their expectations of themselves and thereby freeing themselves of the strictures of intensive motherhood. Instead, the emphasis needs to shift from notions of individual responsibility to analyse and contest dominant discourses of motherhood—this can only be done collectively, which is what the support may have provided for our participants. Indeed, Swale's (2019) recent dissertation highlights how through social support women can "create a village" through which they share experiences and negotiate and reconstruct what it means to be a "good mother" within women's individual contexts. And yet while social support and connection may be a step towards resisting and transforming intensive mothering ideologies, there is a need to engage in societal critique of Eurocentric discourses to consider what the implications of these are, how they constrain women's 'choices,' and how relevant they are to women across societies. In the context of Aotearoa New Zealand, drawing on Māori constructs of children as integrally valued as extensions of whakapapa, and Māori practices of child-rearing as embedded in collective, relational processes with whānau (extended family) forming a protective, intergenerational, and flexibly oriented support network around parents (Le Grice et al., 2017), may offer a way of challenging and resisting these particular individualised and intensive parenting discourses.

Conclusion

Although the number of women who do not have children has increased in recent decades, motherhood remains an expected life trajectory for the majority, including in Aotearoa New Zealand. At the same time, particularly in heteronormative, middle-class contexts, there has been a shift to what has been described as 'intensive mothering': an individualised form of mothering prioritising the needs of children above those of mothers, requiring significant investment, and reflecting a belief in parental determinism, thereby placing significant burden on mothers for how their children develop. Reproductive decision-making and women's experiences do not occur in a vacuum but against the backdrop of such mothering ideologies. Our research highlights the significant impacts these dominant ideologies surrounding parenting can have including on who is encouraged to mother (middle class, resourced mothers) and who is not (e.g., marginalised women). It also highlights how women's reproductive 'choices,' such as planning for parenting, may be constrained by these discourses, and how they may affect the experiences of women with infertility and women's adjustment to parenting. Women may expect themselves to meet particular personal, relational, and financial milestones to position themselves to be 'good' mothers, able to provide for children in such a way as to assure positive outcomes. These expectations may not only delay women's parenthood plans, potentially impacting their fertility, but, in the event of infertility, may also lead women to assume responsibility for their difficulties and struggle to negotiate an identity as a woman without children. Finally, the intensive mothering ideology may also complicate adjustment to parenting, especially after infertility, as women may continue to set high expectations for themselves in relation to their mothering, with potential implications for their wellbeing as parents. Alongside challenging the implicit assumption of these discourses of motherhood as the main life path for women, as this has substantial consequences for those who seek to have children but are unable to do so, our research reflects a need to challenge assumptions of intensive parenting and the expectation that women must reproduce as a matter of public duty. Finally, we need to acknowledge the individualistic and neoliberal assumptions that underlie such ideologies and counterbalance these with other iterations of parenthood available in society, such as those discussed in the chapters throughout this volume, which offer more inclusive ways of recognising diverse kinds of intimate and reproductive citizenship.

References

Allan, H., & Finnerty, G. (2007). The practice gap in the care of women following successful infertility treatments: Unasked research questions in midwifery and nursing. *Human Fertility, 10*(2), 99–104.

Allan, H. T., van den Akker, O., Culley, L., Mounce, G., Odelius, A., & Symon, A. (2019). An integrative literature review of psychosocial factors in the transition to parenthood following non-donor-assisted reproduction compared with spontaneously conceiving couples. *Human Fertility.* https://doi.org/10.1080/14647273.2019.1640901. Epub 2019 July 22.

Bass, B. C. (2015). Preparing for parenthood? Gender, aspirations, and the reproduction of labor market inequality. *Gender & Society, 29*(3), 362–385.

Bell, K. (2013). Constructions of 'infertility' and some lived experiences of involuntary childlessness. *Affilia, 28*(3), 284–295.

Christopher, K. (2012). Extensive mothering: Employed mothers' constructions of the good mother. *Gender & Society, 26*(1), 73–96.

Cooke, A., Mills, T. A., & Lavender, T. (2012). Advanced maternal age: Delayed childbearing is rarely a conscious choice: A qualitative study of women's views and experiences. *International Journal of Nursing Studies, 49*(1), 30–39.

Coombe, J., Loxton, D., Tooth, L., & Byles, J. (2019). 'I can be a mum or a professional, but not both': What women say about their experiences of juggling paid employment with motherhood. *Australian Journal of Social Issues, 54*(3), 305–322.

Correll, A., Benard, S., & Paik, I. (2014). Getting a job: Is there a motherhood penalty? *American Journal of Sociology, 112*(5), 1297–1339.

Daniluk, J. C. (2015). 'Sleepwalking into infertility': The need for a gentle wake-up call. *The American Journal of Bioethics, 15*(11), 52–54. https://doi.org/10.1080/15265161.2015.1088990

DeSouza, R. (2013). Who is a 'good' mother?: Moving beyond individual mothering to examine how mothers are produced historically and socially. *Australian Journal of Child and Family Health Nursing, 10*(2), 15–18.

Donath, O. (2015). Choosing motherhood? Agency and regret within reproduction and mothering retrospective accounts. *Women's Studies International Forum, 53,* 200–209.

Dotti Sani, G., & Treas, J. (2016). Educational gradients in parents' child-care time across countries. *Journal of Marriage and Family, 78*(4), 1083–1096.

Elliott, S., Powell, R., & Brenton, J. (2015). Being a good mom: Low-income, black single mothers negotiate intensive mothering. *Journal of Family Issues, 36*(3), 351–370.

Fisher, J. R., Hammarberg, K., & Baker, H. G. (2005). Assisted conception is a risk factor for postnatal mood disturbance and early parenting difficulties. *Fertility and Sterility, 84*(2), 426–430.

Fisher, J. R., Hammarberg, K., & Baker, G. H. (2008). Antenatal mood and fetal attachment after assisted conception. *Fertility and Sterility, 89*(5), 1103–1112.

Gibson, F. L., Ungerer, J. A., Tennant, C. C., & Saunders, D. M. (2000). Parental adjustment and attitudes to parenting after in vitro fertilization. *Fertility and Sterility, 73*(3), 565–574.

Giesselmann, M., Hagen, M., & Schunck, R. (2018). Motherhood and mental wellbeing in Germany: Linking a longitudinal life course design and the gender perspective on motherhood. *Advances in Life Course Research, 37*, 31–41.

Gotlib, A. (2016). 'But you would be the best mother': Unwomen, Counterstories, and the motherhood mandate. *Journal of Bioethical Inquiry, 13*(2), 327–347.

Graham, A. W. (2018). *Tika Tonu: Young Māori mothers' experiences of wellbeing surrounding the birth of their first tamaiti* (PhD Thesis), Victoria University of Wellington. http://researcharchive.vuw.ac.nz/bitstream/handle/10063/7670/thesis_access.pdf?sequence=1

Gunderson, J., & Barrett, A. E. (2017). Emotional cost of emotional support? The association between intensive mothering and psychological well-being in midlife. *Journal of Family Issues, 38*(7), 992–1009.

Hanser, A., & Li, J. (2017). The hard work of feeding the baby: Breastfeeding and intensive mothering in contemporary urban China. *The Journal of Chinese Sociology, 4*(1), article 18.

Hays, S. (1996). *The cultural contradictions of motherhood*. Yale University Press.

Henderson, A. C., Harmon, S. M., & Houser, J. (2010). A new state of surveillance? Applying Michel Foucault to modern motherhood. *Surveillance & Society, 7*(3/4), 231–247.

Henderson, A., Harmon, S., & Newman, H. (2016). The Price mothers pay, even when they are not buying it: Mental health consequences of idealized motherhood. *Sex Roles, 74*(11–12), 512–526.

Hjelmstedt, A., Widström, A. M., Wramsby, H., & Collins, A. (2004). Emotional adaptation following successful in vitro fertilization. *Fertility and Sterility, 81*(5), 1254–1264.

Ishizuka, P. (2019). Social class, gender, and contemporary parenting standards in the United States: Evidence form a national survey experiment. *Social Forces, 98*(1), 31–58.

Lareau, A. (2011). *Unequal childhoods: Class, race and family life*. University of California Press.

Lavender, T., Logan, J., Cooke, A., Lavender, R., & Mills, T. A. (2015). 'Nature makes you blind to the risks': An exploration of women's views surrounding decisions on the timing of childbearing in contemporary society. *Sexual & Reproductive Healthcare, 6*(3), 157–163.

Le Grice, J., Braun, V., & Wetherell, M. (2017). 'What I reckon is, is that like the love you give to your kids they'll give to someone else and so on and so on': Whanaungatanga and mātauranga Māori in practice. *New Zealand Journal of Psychology, 46*(3), 88–97.

Lee, M., Schoppe-Sullivan, S., & Kamp Dush, C. (2012). Parenting perfectionism and parental adjustment. *Personality and Individual Differences, 52*(3), 454–457.

Letherby, G. (2002). Challenging dominant discourses: Identity and change and the experience of 'infertility' and 'involuntary childlessness'. *Journal of Gender Studies, 11*(3), 277–288.

Lucas, N., Rosario, R., & Shelling, A. (2015). New Zealand university students' knowledge of fertility decline in women via natural pregnancy and assisted reproductive technologies. *Human Fertility, 18*(3), 208–214.

Mackendrick, N. (2014). More work for mother: Chemical body burdens as a maternal responsibility. *Gender & Society, 28*(5), 705–728.

Maehara, K., Iwata, H., Kosaka, M., Kimura, K., & Mori, E. (2020). Experiences of transition to motherhood among pregnant women following assisted reproductive technology: A systematic review protocol of qualitative evidence. *JBI Evidence Synthesis, 18*(1), 74–80.

Maher, J., & Saugeres, L. (2007). To be or not to be a mother?: Women negotiating cultural representations of mothering. *Journal of Sociology, 43*(1), 5–21.

Malik, S. H., & Coulson, N. S. (2008). Computer-mediated infertility support groups: An exploratory study of online experiences. *Patient Education and Counseling, 73*(1), 105–113.

Martin, L. J. (2017). Pushing for the perfect time: Social and biological fertility. *Women's Studies International Forum, 62*, 91–98.

Matley, D. (2020). 'I miss my old life': Regretting motherhood on Mumsnet. *Discourse, Context & Media, 37*, 1–8.

Matthews, A. M., & Matthews, R. (1986). Beyond the mechanics of infertility: Perspectives on the social psychology of infertility and involuntary childlessness. *Family Relations, 35*(4), 479–487.

McLeod, C., & Ponesse, J. (2008). Infertility and moral luck: The politics of women blaming themselves for infertility. *International Journal of Feminist Approaches to Bioethics, 1*(1), 126–144.

McMahon, C. A., Ungerer, J. A., Tennant, C., & Saunders, D. (1997a). Psychosocial adjustment and the quality of the mother-child relationship at four months postpartum after conception by in vitro fertilization. *Fertility and Sterility, 68*(3), 492–500.

McMahon, C. A., Ungerer, J. A., Beaurepaire, J., Tennant, C., & Saunders, D. (1997b). Anxiety during pregnancy and fetal attachment after in-vitro fertilization conception. *Human Reproduction, 12*(1), 176–182.

Mohammadi, N., Shamshiri, M., Mohammadpour, A., Vehvilainen-Julkunen, K., Abbasi, M., & Sadeghi, T. (2014). 'Super-mothers': The meaning of mothering after assisted reproductive technology. *Journal of Reproductive and Infant Psychology, 33*(1), 42–53.

Nomaguchi, K., & Milkie, M. (2020). Parenthood and wellbeing: A decade in review. *Journal of Marriage and Family, 82*, 198–223.

Payne, D., & Goedeke, S. (2009). *Enforcing motherhood: Assisted reproductive technologies and the media.* 15th international critical and feminist perspectives in Health & Social Justice Conference. April 16–19, AUT University, Auckland.

Pedersen, S. (2016). The good, the bad and the 'good enough' mother on the UK parenting forum Mumsnet. *Women's Studies International Forum, 59*, 32–38.

Perrier, M. (2013). No right time: The significance of reproductive timing for younger and older mothers' moralities. *The Sociological Review, 61*(1), 69–87.

Petersen, E. (2015). Mommy bloggers as rebels and community builders: A generic description. *Journal of the Motherhood Initiative for Research and Community Involvement, 6*(1), 9–30.

Peterson, B. D., Pirritano, M., Tucker, L., & Lampic, C. (2012). Fertility awareness and parenting attitudes among American male and female undergraduate university students. *Human Reproduction, 27*(5), 1375–1382.

Shreffler, K., Greil, A., Tiemeyer, S., & McQuillan, J. (2020). Is infertility resolution associated with a change in women's well-being? *Human Reproduction, 35*(3), 605–616.

Statistics New Zealand. (2018). Parenting and fertility trends in New Zealand 2018. https://www.stats.govt.nz/reports/parenting-and-fertility-trends-in-new-zealand-2018

Swale, M. (2019). *'I am actually doing alright': A grounded theory exploration of how women's online social support use affects maternal identity construction and wellbeing* (MA Thesis), Massey University, Palmerston North. https://mrons.massey.ac.nz/handle/10179/15663

Tonkin, L. (2019). *Motherhood missed: Stories from women who are childless by circumstance.* Jessica Kingsley.

Turnbull, B., Graham, M., & Taket, A. (2017). Pronatalism and social exclusion in Australian society: Experiences of women in their reproductive years with no children. *Gender Issues, 34,* 333–354.

Ulrich, M., & Weatherall, A. (2000). Motherhood and infertility: Viewing motherhood through the lens of infertility. *Feminism & Psychology, 10*(3), 323–336.

Ussher, S. R. (2015). *Women and careers: New Zealand women's engagement in career and family planning* (MA Thesis), University of Waikato, Hamilton, Waikato. http://hdl.handle.net/10289/9370

Wirtberg, I., Möller, A., Hogström, L., Tronstad, S. E., & Lalos, A. (2007). Life 20 years after unsuccessful infertility treatment. *Human Reproduction, 22*(2), 598–604.

Index[1]

A
Ableism, 161
Abortion, 32, 38, 46, 245, 247, 303
Access to fertility treatment, 7, 10, 12, 157, 173
Adoption, 3, 16, 76, 87, 91, 108, 128, 132, 134, 136, 144, 179, 204, 209, 214, 215, 221, 224n4, 231, 234, 259, 271
Adrian, Stine, 13, 34, 47
Advisory Committee on Assisted Reproductive Technology (ACART), 17, 180, 206
Affect, 6, 20, 33, 36, 41, 46, 47, 52, 57, 59, 60, 79, 85, 89, 90, 95n1, 104, 112, 129, 133, 138, 159–161, 222, 265, 289, 291, 295, 304, 309, 318, 320, 323
Affective, 31–47, 51–70, 85, 87–89, 94, 107, 118, 133, 154, 160, 161, 168, 242
Affect theory, 52, 161
Ageism, 161
Age-related fertility, 4
Ahmed, Sara, 33, 47, 52, 57, 69
Altruistic, 13, 18, 229
Animacy, 31–47
Animacy hierarchies, 14, 36, 40–42
Anonymity, 20, 181, 183, 185, 282, 284, 289
Aotearoa, 8, 12, 13, 18, 19, 21n3, 21n4, 21n6, 155–157, 253, 254, 260, 274n1
Aotearoa New Zealand, 7, 17–19, 21n3, 96n1, 96n5, 153–174, 229–250, 253, 264, 303–306, 311, 312, 322

[1] Note: Page numbers followed by 'n' refer to notes.

Index

Assisted reproduction, 3, 10, 12, 18, 20, 54, 101, 102, 113–115, 131, 155–157, 171–174, 203, 205, 212, 221, 222, 229–250, 253, 256, 320

Assisted reproductive technologies (ARTs), vi, 2–4, 6, 7, 9–12, 14, 16, 17, 20, 21n5, 38, 43, 44, 46, 75, 90, 128, 154, 203, 208, 223, 224, 231, 253, 283, 288, 289, 296, 311, 313, 319

Australia, 12, 76, 80, 109, 181–183, 195, 204, 280, 282, 283, 285, 303

B

Bacchi, Carol, 159, 160
Baldwin, Kylie, 4, 5, 107, 108, 110, 111
Bell, Ann, 3, 6, 7, 10
Berend, Zsuzsa, 231, 237, 238, 240–242, 244, 246
Bio-banking, 110, 113, 115
Bioeconomy, 55, 104
Biological citizenship, 1
Biological clock, 42, 43, 101
Biological relatedness, 18, 77, 91, 193
Biopolitics, 2, 6
Biopreparedness, 118
Birth certificate, 17, 268, 290
Birth mother, 204
Body Mass Index (BMI), 7, 17, 32, 154–157, 159, 160, 162, 163, 165–167, 171, 172
Breast cancer, 64
Briggs, Laura, 3, 11

C

Canada, 80, 155, 264
Cancer, 4, 8, 10, 51, 58–64
Cancer treatment, 51, 55, 62, 68, 70, 238
Capitalist, 104, 110, 112, 116
Chemotherapy, 60, 106
Childfree, 20, 43, 313, 314, 316, 317
Choice, 3, 4, 7–9, 15, 16, 19, 57, 61, 62, 67, 69, 76, 84, 85, 102, 106–110, 112–114, 118, 131, 135–138, 143, 144, 154, 157, 163, 166, 172, 179–197, 240, 250, 254, 256, 257, 267, 271–273, 281, 284, 293, 303–323
Choice Mothers, 17, 179, 180, 182–197
Chrononormativity, 43
Cisgender, 6, 9, 10, 76, 77, 79, 87, 159
Cisnormative, 82, 89–94
Cisnormativity, 76, 90, 94
Citizenship, vi, vii, 1–21, 44, 77, 116, 138, 145, 204, 323
Class, 11, 21n5, 136, 155, 166, 173, 304–306, 323
Clinical Priority Assessment Criteria (CPAC), 7, 8, 17, 154–157, 160, 169, 173
Colonialism, 167
Commercialisation, 7
Compensation, 13, 230
Content analysis, 80
Convention of the Rights of the Child, 296
Cryobanks, 52

Cryopreservation, 4, 5, 8, 14, 15, 32, 34, 36, 46, 52, 53, 55, 56, 58, 59, 69, 80, 95n1, 102, 106, 107, 109
Culley, Lorraine, 108, 110, 111
Cultural imaginaries, 37, 43

D
Danish, 14, 31–47, 52–54, 58
Dating rules, 247
De Lacy, Sheryl, 6, 8
Dempsey, Deborah, 10, 12, 183, 184, 186–189, 254, 270, 283
Denmark, 12, 13, 31, 32, 35, 38, 40, 43, 45, 53–54, 58, 59, 70
Depression, 161, 168, 318
Divorce, 262–264, 266, 269, 272
DNA matching, 285
Donation, 13, 35, 37, 45, 55, 68, 180, 181, 188–190, 194, 207, 208, 254, 256, 265, 270, 282, 283
Donor-conceived people, 17, 20, 181, 197, 280–292, 294, 296
Donor-conceived person, 20, 280, 281, 290, 291, 293
Donor conception, 17, 20, 182, 184, 185, 194, 279–296
Donor insemination, 179, 180, 195, 231, 254, 258, 270, 284
Donor-linking, 12, 183, 186, 192
Donor siblings, 17, 179–197, 283, 285, 286
Due, Clemence, 5, 138, 230, 249
Dysphoria, 90–92, 94

E
Economic marginalisation, 161
Edelman, Lee, 4, 60
Egg retrieval, 54
Eggs, 8, 10–15, 21n4, 31, 32, 38, 45, 46, 51, 53–57, 62, 64, 66–69, 75, 76, 80, 83, 86, 87, 90, 95n1, 101–119, 129, 130, 137, 204, 207, 214, 217, 219, 220, 224n1, 241
Elective egg freezing, 102
Embodiment, 16, 37, 77, 129, 133, 135, 139–145, 167, 174
Embryo freezing, 14, 69
England, 204
Epistemic community, 79, 84
Ethics Committee on Assisted Reproductive Technology (ECART), 205, 206, 208, 230, 231, 234
Ethnicity, 11, 80, 166, 184, 212, 306
Ethnographic, 58, 104, 242
Ethnographic fieldwork, 19, 231
Eugenicist, 159, 173
Experiential expert, 15, 79, 83–85
Experiential knowledge, 83, 84, 92, 94

F
Facebook, 35, 108, 281, 282, 285, 286, 291
Family building, 5, 19, 75, 77, 78, 91, 132, 180, 193
Family life, 43, 57, 63, 263, 312
Fatherhood, 19, 90, 253–273
Fat shaming, 165, 167, 168, 173

332 Index

Fat women, 154, 159, 292
Fertility clinics, 6, 8, 18, 21n4, 34, 35, 60, 61, 64, 77, 91, 94, 96n5, 114, 168, 180, 181, 183, 185–187, 189, 204, 205, 229, 231, 254, 288
Fertility counsellors, 245
Fertility preservation, 4, 8, 12, 14, 15, 51–70, 75–95, 101–103, 105, 107, 108, 112, 114, 115, 118, 119
Fertility specialists, 54, 55, 155
Foetal, 56, 158
Foucault, Michel, 2
France, 204
Franklin, Sarah, 37, 41, 52, 57, 58, 68, 128, 130–132, 138
Fraser, Nancy, 118
Freeman, Elizabeth, 4, 34, 42, 43
Freezing time, 15
Friend, 35, 36, 43, 66, 84, 118, 181, 186, 231, 235, 249, 253, 255, 267, 287, 291, 307, 310
Friendship, 237, 238
Future infertility, 52
Futurity, 14, 20, 34, 56, 60–63, 85–87, 118, 119

G

Gamete freezing, 14, 51, 75
Gamete storage, 81, 88, 89, 91, 93, 94, 95n1
Gay couple, 19, 232, 240, 254, 259, 260, 271
Gay fatherhood, 19, 253–273
Gender-affirming hormone therapy, 8, 75, 76

Genealogy, 181, 182, 214, 224n3
General practitioner (GP), 155, 160, 162, 163, 165, 174n3
Generation, 20, 56, 63, 167, 283, 287, 288, 304
Generational time, 44
Genetic connection, 209, 215, 217, 218, 220
Genetic link, 18, 135, 203–224, 261
Germany, 80, 204
Gestational surrogacy, 13, 18, 224n1, 229–232, 237, 245, 249, 250, 259, 283
Gift, 104, 190, 208, 317
Golombok, Susan, 2, 3, 182, 185, 194
Good citizen, vi, 4, 169
Good mothers, 5, 20, 304, 306–313, 315–316, 322, 323
Greece, 204
Greil, Arthur, 3, 6, 161
Grief, 161, 260, 313, 314

H

Halberstam, Jack, 4, 42–44
Hays, Sharon, 304, 309
Healthcare providers, 76, 77, 94
Heterosex, 5, 9, 230
Heterosexism, 161
Heterosexual couples, 6, 32, 180, 182, 266, 270
Homonormativity, 3, 4
Hope, vi, 14, 16, 21, 51–70, 84, 108, 115, 129, 233, 243, 248, 268, 281, 295
Hope technology, 52, 57, 69

Hormones, 4, 8, 53, 75, 76, 84, 86–89, 91, 95n1, 96n3, 104
Hudson, Nicky, 55, 129, 133, 142
Human Assisted Reproductive Technology Act (HART Act), 19, 180, 185, 192, 204, 205, 207, 254
Human right, 204, 287, 290
Hysterectomy, 130, 238

I

Immunosuppressive therapy, 130
Indigenous, 153, 158, 159, 169, 173, 174n4, 181, 305
Infertility, vi, 1–21, 44, 52, 55, 70, 101, 128, 131, 134, 138, 139, 144, 145, 154, 155, 157, 158, 160–165, 167–171, 214, 230, 282, 303–323
Informed consent, 8, 36, 81, 82, 93
Inhorn, Marcia C., 10, 12, 21n5, 55–57, 69, 91
Insurance, 10, 51, 62, 77, 106–108, 111, 112, 115
Intended parents (IPs), 13, 19, 203–205, 207–209, 213, 215–223, 230–232, 234–250, 250n1
Intensive mothering, 117, 304, 306, 307, 309, 314, 317, 321–323
Intersectional, 167, 175n6, 281
Intersectionality, 160, 161, 174n6
Interviews, 14, 16, 17, 19, 34–36, 39–42, 45, 51, 52, 58–60, 63, 66, 104, 108, 110, 111, 132, 136, 154, 159–161, 184, 222, 231, 235, 254, 255, 258, 267, 269, 291–294, 305

Intimacies, 15, 87–89, 131, 291
Intimate citizenship, 1, 3, 323
Intimate relationships, 5, 15, 138, 233, 263
In vitro fertilisation (IVF), 16, 31, 34, 35, 38, 44, 46, 52, 54, 64, 128–130, 146n5, 203, 214, 224n1, 283, 313, 319, 322
Involuntary childlessness, 3, 59
IVF surrogacy, 203, 224n1

K

Kinning, 34, 37, 40–42, 47
Kinship, 3, 5, 14, 18, 34, 40, 63, 70, 90, 105, 181, 194, 224n3, 230, 233, 242, 244, 250, 253, 255–258
Known donor, 192, 253–273
Kroløkke, Charlotte, 4, 34, 42, 44, 51, 52, 55, 56, 62, 107, 111, 132

L

Latent life, 32, 39
Lawyer, 245, 247, 249
Legal parentage, 204, 215–216
Legal parents, 203, 204, 215, 216, 222, 231, 266, 274n1
Lesbian, gay, bisexual, transgender, and queer (LGBTQ), 2, 4, 6, 9, 11, 77
Lesbian known donor reproduction, 19
Lesbian mothers, 182
Letter of intent, 245–247, 249, 250
Liberal feminism, 106, 116

M

Mamo, Laura, 4, 18, 77, 131
Māori, 21n3, 156, 160, 166, 167, 173, 174n4, 175n7, 181, 208, 212, 224n3, 224n4, 306, 307, 312, 322
Medical egg freezing, 10, 57, 105, 106
Medicalisation, 37
Medical transition, 4, 96n3
Menopause, 7, 53, 55, 66–68, 70, 102
Menstrual cycle, 53, 66–68, 70
Middle-class, 103, 117, 175n6, 304–306, 308, 322, 323
Misgendering, 91, 92
Mommy blogs, 321
Motherhood, 14, 15, 20, 57, 90, 102, 104, 106, 107, 110–112, 117–119, 142, 180, 269, 273, 303–319, 321–323

N

Neoliberalism, 5, 11, 15, 105, 116, 117
Netherlands, 181
New Zealand, 7, 19, 20, 21n3, 77, 80, 87, 93, 160, 166, 174n3, 179–197, 204–207, 210, 223, 224n1, 224n3, 224n4, 229–236, 240, 242, 246, 247, 249, 250, 283
Non-binary people, 3, 4, 15, 21n1, 104
Nordqvist, Petra, 18, 19, 131, 181, 255, 256, 270
Norway, 181

O

Obese, 156, 168, 172
Obesity, 17, 156–159, 174n1
Oestrogen, 84, 87
Online support network, 5
Ontological choreography, 33
Oöcytes, 14, 21n4, 31, 32, 38, 41, 43, 45, 62, 102, 103, 106, 107, 109, 111, 130, 283
Open-identity donor, 181, 185, 197
Organ transplantation, 130
Ovarian egg freezing, 13, 95n1
Ovarian tissue freezing, 14, 51–70, 113

P

Pacific, 156, 160, 166, 167, 173
Pākehā, 184, 231, 305–307
Parenthood, 3, 4, 9, 11, 16, 20, 56, 69, 76–78, 93, 105, 128, 131, 134, 135, 137, 143, 173, 179, 180, 203, 223, 258–273, 274n1, 303–323
Pearce, Ruth, 76, 77, 79, 84, 92
Plummer, Ken, 1, 3, 4
Portugal, 204
Positionality, 79, 280, 281, 291
Postmenopausal, 102
Potentially infertile, 56, 102
Potentially maternal, 101–119
Pregnancy, 6, 8, 18, 32, 35, 37, 39, 46, 59, 63, 64, 69, 90, 128–131, 134, 136, 137, 144, 145, 146n5, 153, 154, 157–159, 165, 167, 169, 172, 190, 203, 205, 206, 216, 218, 220, 239, 244, 311, 318–320

Pregnant, 8, 10, 31, 41, 44, 45, 59, 65, 90, 107, 111, 113, 118, 130, 134, 157, 158, 162, 163, 165–167, 169, 170, 172, 174n2, 239, 245, 247, 269
Pronatalist, 105, 303, 313, 317
Psychoanalytic, 104, 118
Public funding, 7, 8
Public healthcare, 12, 54, 61

Q

Qualitative interviews, 14, 19, 52, 104, 154, 159, 254, 293
Qualitative researchers, 280

R

Race, 155, 173
Racialisation, 37, 104, 168, 173
Radin, Joanna, 32, 52, 55, 56, 111
Raun, Tobias, 78, 80, 87, 89, 93
Reasonable expenses, 13
Reflexivity, 279–296
Relatedness, 18, 19, 34, 77, 91, 105, 193, 208, 254–256, 258, 260, 261, 263
Relationality, 255, 271
Reproductive autonomy, 20, 32, 60–63, 69, 104, 131, 159, 303, 306
Reproductive citizens, 5, 6, 18, 44
Reproductive citizenship, vi, 1–21, 138, 323
Reproductive futures, 4, 16, 75, 78, 79, 82, 86, 87, 90, 92–94, 144
Reproductive justice, 17, 104, 153–174

Reproductive liberty, 16, 129, 137, 138
Reproductive vulnerability, 5, 11, 19, 20, 230
Responsibilising, 158
Responsibilization, 11
Riggs, Damian, 5, 76–78, 84, 91–93, 105, 118, 138, 230, 249, 254
Risk management, 52, 63, 69, 115
Romance, 234, 242, 244, 250, 250n2
Rose, Nikolas, 1, 2, 11, 44, 52

S

Sandelowski, Margarete, 6, 8, 134, 144
Secrecy, 20, 194, 282, 284, 289, 295
Self-blame, 161, 163–166, 168
Semi-structured interviews, 17, 34, 132, 184, 231, 305
Sexual citizenship, 2
Sexually transmitted infections, 7
Shadow-legalities, 232–236, 242, 245, 247–250
Shame, 158, 161, 165–168, 282, 294, 321
Single Mothers by Choice (SMC), 2, 17, 179–197
Single parent, 180
Situational fertility barriers, 9, 11
Sizeism, 161
Smart, Carol, 255–257
Social egg freezing, 10, 56, 101–119
Social imaginaries, 129, 132–145

Social media, 8, 35, 58, 80–82, 88, 160, 231, 253, 285, 296, 304, 310
Sociology of personal life, 19, 255–257
South Africa, 204
Spain, 204
Sperm donor, 17, 179–197, 214, 216, 270
Sperm freezing, 95n1
Stigma, 76, 131, 165–168, 304, 311, 317, 321
Storytelling, 258
Stratified reproduction, 10
Structural infertility, vi, 10–11, 13, 17, 21
Surrogacy, 5, 13, 16–19, 44, 87, 91, 104, 108, 116, 119n1, 128, 132, 134, 136, 137, 141, 144, 145n5, 203–224, 229–250, 259, 283
Survey, 18, 76, 114, 205, 210–211, 215–219, 221–224, 224n2
Sweden, 12, 13, 16, 128, 145n5, 181
Swedish, 13, 44, 128–130, 132
Synchronisation, 42–45, 47, 66–68, 70

T

Taiwan, 204
Teman, Elly, 240, 241
Temporality/temporalities, 14, 31–47, 51–70
Testosterone, 86, 87
Thailand, 204
Thompson, Charis, 33, 37, 39, 47, 57, 128, 130, 131, 142

Time, 4, 9, 15, 19, 34, 35, 39, 40, 42–47, 51–70, 81, 86, 88, 89, 91, 96n1, 101, 102, 106–108, 110–117, 119, 130, 135, 136, 141, 156, 168, 170, 171, 181, 184, 187, 189, 194, 196, 203, 207, 211, 223, 231, 232, 234, 239, 240, 242, 243, 247–249, 254, 261, 262, 266, 267, 269, 270, 272, 273, 281, 282, 286, 287, 294, 296, 304–308, 311, 312, 316, 319–321
Tissue economy, 104
Traditional surrogacy, 18, 224n1, 229–232, 237, 238, 240, 244, 245, 250
Transfeminine, 87, 88, 90
Transgender, 3, 4, 8, 12, 15, 21n1, 21n4, 75, 80, 96n2, 104, 105, 222
Transition, 4, 75, 78, 86, 87, 92, 93, 96n3, 270, 286, 313, 317, 318
Transmasculine, 88, 89
Transphobia, 76
Turner, Bryan, 2, 3, 5, 16, 77, 182, 189, 194

U

Umbilical cord blood, 55
Uncle, 255, 262, 263
United Kingdom, 12, 80, 155, 181
United States (USA), 7, 11, 12, 80, 231, 237, 240–242, 246, 288
Uterine factor infertility (UFI), 128, 136, 138
Uterus, 16, 31, 113, 127–145

Uterus transplantation (UTx), 16, 127–145
UTx-IVF, 16, 128–137, 139–145

V

Vagina, 91, 128, 130, 140, 141, 143
Van Balen, Frank, 10, 12
Van de Wiel, Lucy, 4, 10, 14, 51, 55, 56, 91, 102, 103
Victoria, Australia, 12, 108, 183, 195
Victorian Assisted Reproductive Treatment Authority (VARTA), 108, 120n4, 183, 192, 195, 196
Video blogs (vlogs), 15, 75–95
Vitrification, 106

W

Waldby, Catherine, 37, 43, 51–53, 55, 62, 70
War on obesity, 156, 157
Weight-based exclusion, 154, 162
Weight-loss, 17, 155–157, 163–165, 170–172
Whakapapa, 167, 175n7, 181, 208, 209, 224n3, 322
Whānau, 167, 175n7, 191, 208, 214, 224n3, 322
World Health Organization (WHO), 6, 288

Y

YouTube, 15, 75–95
YouTube copyright, 81